ALTERNATIVE PLANNING HISTORY AND THEORY

This book includes 12 newly commissioned and carefully curated chapters each of which presents an alternative planning history and theory written from the perspective of groups that have been historically marginalized or neglected.

In teaching planning history and theory, many planning programs tend to follow the planning cannon – a normative perspective that mostly accounts for the experience of white, Anglo, Christian, middle class, middle aged, heterosexual, able-bodied, men. This book takes a unique approach. It provides alternative planning history and theory timelines for each of the following groups: women, the poor, LGBTQ+ communities, people with disabilities, older adults, children, religious minorities, people of color, migrants, Indigenous people, and colonized peoples (in South Asia and Sub-Saharan Anglophone Africa). To allow for easy cross-comparison, chapters follow a similar chronological structure, which extends from the late 19th century into the present. The authors provide insights into the core planning issues in each time period, and review the different stances and critiques.

The book is a must-read for planning students and instructors. Each chapter includes the following pedagogical features: (1) a boxed case study which presents a recent example of positive change to showcase theory in practice; (2) a table which lays out an alternative planning history and theory timeline for the group covered in the chapter; and (3) suggestions for further study comprising non-academic sources such as books, websites, and films.

Dorina Pojani is Associate Professor in urban planning at the University of Queensland, Australia. She teaches planning theory and history. Her research is international and comparative and covers a variety of built environment topics, including urban design, transport, and housing.

ALTERNATIVE PLANNING HISTORY AND THEORY

Edited by Dorina Pojani

Routledge
Taylor & Francis Group

LONDON AND NEW YORK

Cover image: Getty Images/Abeltx

First published 2023
by Routledge
4 Park Square, Milton Park, Abingdon, Oxon OX14 4RN

and by Routledge
605 Third Avenue, New York, NY 10158

Routledge is an imprint of the Taylor & Francis Group, an informa business

British Library Cataloguing-in-Publication Data
A catalogue record for this book is available from the British Library

ISBN: 9780367743888 (hbk)
ISBN: 9780367743895 (pbk)
ISBN: 9781003157588 (ebk)

DOI: 10.4324/9781003157588

Typeset in Bembo
by KnowledgeWorks Global Ltd.

Dedication

To my parents, who have bravely endured Communism, Capitalism, and the Patriarchy.

- Dorina Pojani

CONTENTS

ACKNOWLEDGMENTS

Many thanks to all of the authors for their invaluable contributions to this volume. My thanks also to Faye Leering from Routledge for guiding the book project from start to finish. Special thanks to my family, which is split between Albania, Australia, and America.

- The editor

CONTRIBUTORS

Sangeeta Banerji is Post-doctoral Research Associate at Brown University. Her research focuses on technological reform within urban bureaucracies and brokerage in Indian cities.

Enrique M. Buelna is a faculty member at Cabrillo College. He focuses on Chicano history and the intersections of class, race, labor, radical activism, civil rights, immigration, culture, and identity. He is the author of *Chicano Communists and the Struggle for Social Justice* (2019) and holds a doctorate in history from the University of California, Irvine.

Gabriel Buelna is an attorney and a faculty member in Chicana/o Studies at Cal State Northridge. He is also a Trustee at the Los Angeles Community College District. His work focuses on issues of critical race theory within Mexican American communities, specifically the causal connections between anti-Mexicanism and representation and resource allocation. He has a YouTube channel called *BuelnaNews* and has written extensively on the 2020 Presidential Election. He holds a doctorate in political science from the Claremont School of Politics and Economics.

Kevin M. Dunn is Pro Vice-Chancellor Research and Professor in human geography and urban studies at Western Sydney University. His research has highlighted the culturally and spatially uneven distribution of citizenship in Australia. He is author of over thirty chapters in books and over seventy articles. His most recent articles are on racism in the sharing economy, anti-racism education and Islamophobia. They are published in *Australian Journal of Social Issues* (2020), *Ethnicities* (2021), and *Geoforum* (2019). He leads the national award winning *Challenging Racism Project*, delivering impactful research that has underpinned national anti-racism strategies.

D. Asher Ghertner is Associate Professor at Rutgers University. He is the author of *Rule by Aesthetics: World-Class City Making in Delhi* (2015) and co-editor of *Futureproof: Security Aesthetics and the Management of Life* (2020) and *Land Fictions: The Commodification of Land in City and Country* (2021). His research focuses on environmental politics and informal housing in India.

Melissa Heil is Assistant Professor at Illinois State University where she teaches courses in urban geography and urban planning. Her current research examines the intersection of racial inequality, austerity, and urban water systems in rust belt cities. Her previous research has investigated the changing role of Detroit's community development corporations and inequitable economic development in Flint, Michigan.

Ella Howard is Associate Professor at Wentworth Institute of Technology, where she teaches urban and digital history. Her book *Homeless: Poverty and Place in Urban America* (2013) traced the rise and fall of New York City's Bowery as a skid row. Her current project analyzes the role of historic preservation in fostering and maintaining economic and racial segregation. She has also published on the history of design and material culture, feminist historiography, and popular culture. She earned a doctorate in American and New England Studies at Boston University.

Álvaro Huerta is Associate Professor in urban and regional planning and ethnic and women's studies at California State Polytechnic University. He is also a Religion & Public Life Organizing Fellow at Harvard Divinity School. Among other publications, he is the author of the books: *Defending Latina/o Immigrant Communities: The Xenophobia Era of Trump and Beyond* (2019) and *Reframing the Latino Immigration Debate: Towards a Humanistic Paradigm* (2013). He holds a Ph.D. in City and Regional Planning from the University of California, Berkeley.

Rhonda Itaoui is a Postdoctoral Research Fellow and Associate Lecturer in human geography at Western Sydney University. Her research has explored geographies of diversity, multiculturalism and belonging in urban spaces in Australia and the United States. Rhonda's recent articles have focused on the geographies of Islamophobia in Sydney, Australia and the San Francisco Bay Area, the United States. They are published in: *Social and Cultural Geography* (2020), *Mobilities* (2021), and the *Journal for Intercultural Studies* (2019). This research has revealed the need for local and context-specific anti-racism policy practice, public education campaigns and policy initiatives that respond to the spatial imaginaries and lived experiences of racialized groups.

Louise C. Johnson is Honorary Professor at Deakin University and Honorary (Professorial) Fellow at the University of Melbourne. As well teaching in Australian and New Zealand universities for over forty years, Louise has a solid record of research, policy development and community activism. A human geographer, she has researched suburban housing, changing manufacturing workplaces and the dynamics of Australian regional economies. Louise has also done research on urban access and inclusion and on the Indigenous absence in Australian planning. In 2011 she received the Institute of Australian Geographers Australia and International Medal for her contribution to urban, social and cultural geography.

Sukanya Krisnamurthy is a Chancellors Fellow/Senior Lecturer at the University of Edinburgh. Her focus lies at the interface between urban and social geography, where her scholarship analyses how cities can use their resources and values for more sustainable development. Key research areas include: children's lived experiences in diverse urban contexts, and participatory methods for engaging children and youth in design and planning.

John Lewis is Associate Professor at the University of Waterloo and a citizen of the Métis Nation of Ontario. He has worked as a community planner and consultant for municipal and First Nations governments in British Columbia and Ontario and as has been an age-friendly planning researcher for the Ontario Ministry of Seniors' Affairs since 2010.

Garth Myers is Director of the Center for Urban and Global Studies and the Paul E. Raether Distinguished Professor of Urban International Studies at Trinity College in Hartford, Connecticut. He focuses on African urban geography and planning, comparative urbanism, and urban political ecology in his research. He has published seven books and more than 80 book chapters and articles on these themes. His primary focus has been with cities in Tanzania, Kenya and Zambia.

Tiffany Muller Myrdahl is a Senior Lecturer at Simon Fraser University (unceded territory of the Musqueam, Squamish, Tsleil-Waututh and Kwikwetlem Nations, Vancouver, Canada). She completed a Master in Public Policy (2002) and PhD in Geography with a Certificate in Feminist Studies (2008) from the University of Minnesota. Her work examines urban inequalities and inclusion strategies, especially those targeting women and LGBTQ+ communities.

Samantha Ngui's Doctorate of Philosophy was awarded from the University of New South Wales. Her thesis entitled: "*Freedom to worship: Frameworks for the realisation of religious minority rights*" examined the experiences of religious groups in establishing places of worship in Sydney. The uneven outcomes, bias and discrimination identified in the processes informed a discussion of multicultural citizenship and secularism.

Francis Owusu is the Chair of the Department of Community and Regional Planning at Iowa State University. He teaches international planning courses, including planning in developing countries, world cities and globalization, economic and urban planning, and planning methods. His research focuses on Africa, especially Ghana, and has published extensively on globalization, development policy, public sector reforms and capacity building, and urban development and livelihood issues.

Meg Parsons is a historical geographer of New Zealand Māori, Lebanese, and Pākehā/European heritage, whose research examines how Indigenous communities in Oceania understand and respond to environmental changes, and the legacies of colonization on Indigenous health and well-being. She is a Senior Lecturer in Human Geography and Environmental Management at the University of Auckland and a contributing author to the Sixth Assessment of the Intergovernmental Panel on Climate Change.

Sean Peacock is a Researcher based at Newcastle University. His research involves designing digital technologies for and with young people to help them improve their local neighborhoods. Conducting action research with schools and youth groups, he has expertise in designing and evaluating digital technologies to support the inclusion of youth in urban planning processes and environmental issues.

Jessica Ellen Sewell is Associate Professor of urban and environmental planning, architecture history, and American studies and Co-Director of the Center for Cultural Landscapes at the University of Virginia. A feminist historian of cities, buildings, and things, she is author of *Women and the Everyday City: Public Space in San Francisco, 1890-1915* and teaches on gender, race, and cities. She is currently working on books on gender in vernacular architecture and on masculinity and the bachelor pad, as well as an online guide to the urban cultural landscapes of Suzhou, China.

Teresa Strachan is Senior Lecturer at Newcastle University and a Chartered Town Planner. She founded the *YES Planning* student volunteer engagement project in 2014 and focuses her teaching and research on engagement skills and methodologies on working with young people.

Richard Tucker has published nearly 100 outputs on sustainable and universal design, urban design, and relationships between health, accessibility and inclusivity in built environment design. Richard is a director of the HOME Research hub – an interdisciplinary group of Deakin researchers which works with local communities to co-design solutions to complex problems of affordable housing, homelessness and social inclusion. Most recently, he was project leader of the *Accessible & Inclusive Geelong Feasibility Study*, which informed a collective plan of action to enable Geelong to be accessed, understood and used to the greatest extent possible by all people.

Lou Turner is Clinical Assistant Professor of urban and regional planning at the University of Illinois and the Black Arts Curriculum Coordinator for the College of Fine & Applied Arts. Formerly, he held the positions of Academic Advisor and Curriculum Coordinator for the Department of African American Studies at the University of Illinois and Assistant Professor of Sociology at North Central College in Naperville, Illinois. Lou Turner was Director of Research & Public Policy for the Developing Communities Project on Chicago's far South Side, where he guided its strategic and public policy work concerning the Chicago Transit Authority Red Line extension project.

Valerie Watchorn is an occupational therapist and lecturer at Deakin University. She is actively involved in research and education relating to environmental design, universal design and assistive technology as enablers to inclusion and participation, in particular for those who experience disability. Valerie's research strongly recognizes the value of involving people with lived experience of disability in the design of built environments and how this can enhance the application of universal design. Valerie has been involved in service delivery, education and advocacy on issues relating to disability for over 20 years.

David Wilson is Professor of geography, urban planning, and the unit for criticism and interpretive theory at the University of Illinois at Urbana-Champaign. His work focuses on the economic and political issues and challenges that face rust belt cities in the United States and Europe. His most recent books are *Chicago's New Racial Redevelopment Machine and South Side*, 2018; *The Handbook on Spaces of Urban Politics*, Routledge, 2018 (with K. Ward, A. Jonas, and B. Miller); and Making Creative Cities: New Inequalities, 2017 (with U. Gerhard and M. Hoelscher).

Jenny Wood is Research Associate at Heriot-Watt University, where she researches homelessness, poverty and severe and multiple disadvantage. She also researches and acts on the spatial implications of the UN Convention of the Rights of the Child, through her Co-founded Scottish charity *A Place in Childhood* (APiC).

PLANNING HISTORY AND THEORY

Enriching the canon

Dorina Pojani

Introduction

In teaching planning history and theory, many planning programs in the Anglosphere[1] (and some former colonies) tend to follow the academic cannon, which is summarized in Table (see page 12). This cannon is correct but incomplete.[2] To fill the gaps in a systematic manner, this book presents planning history and theory from the perspective of the following groups:

1. Women
2. The poor
3. LGBTQ+ communities
4. People with disabilities
5. Older adults
6. Children
7. Religious minorities
8. People of color
9. Migrants
10. Indigenous people
11. Colonized peoples (Indian Subcontinent)
12. Colonized peoples (Sub-Saharan Anglophone Africa)

As seen in the Table of Contents, each chapter is dedicated to one group.

Theoretical underpinnings

A postmodern approach?

At first sight, this book's approach to planning history and theory appears to be postmodern. Postmodernism is a mode of thought characterized by plurality, social constructivism, contradiction, and a preoccupation with power – and these

DOI: 10.4324/9781003157588-1

features are certainly present in the book.[3] The authors consider planning discourses and practices as inherently ideological – and post-modernism, as a movement, is acutely sensitive to the role of ideology in asserting and maintaining power.[4] A recurring theme in planning theory is that megaprojects are built and zoning practices are adopted which serve – primarily or solely – the interests of the powerful while all "others" are excluded from participating in the process or benefiting from the outcomes. Public participation processes tend to be dominated by groups which hold more power – and often exercise it against others. This produces disenfranchisement and disparate outcomes in the city.

In accordance with postmodernism, a majority of this book's authors accept that many of the identities represented here are, at least in part, socially constructed rather than innate. For example, age is based on objective reality but the views of the roles and capabilities associated with children and older adults are based on social expectations and prejudices. Race – a concept based on skin color and superficial physical traits – is also a social construct which emerged in conjunction with, and as a justification for, colonialism. Urban inequality is an outcome of a market economy which distributes advantages and disadvantages in an unfair manner. But it may also be considered as an identity, especially in the case of intergenerational poverty, which overlaps with social class.

The authors assume that the people belonging in certain identity groups – while being unique individuals – share some of the same experiences and perceptions of the city due to their similar position in the social structure. Moreover, they face injustices based on lingering prejudices about the group to which they belong rather than being recognized as individuals. Conversely, a grouping by identity makes sense because much political action has been achieved on the basis of group empowerment. Movements such as feminism, Civil Rights, Indigenous people's rights, disability rights, gay and lesbian rights, and so on, have relied on group solidarity and support. Also, identity groups often fulfill a human need for comfort and belonging. While some groups have been forced by the planning systems to segregate from others, self-clustering of minorities – e.g., some LGBTQ+ and migrant communities – is common too, as people prefer to live close to those who share their culture and interests.

The contributors to this book clearly advocate on behalf of people and groups with less or no power. The authors examine some of the attitudes and narratives that have caused economic and psychological discomfort to marginalized people over the decades. They also seek to reveal hidden biases and underexamined assumptions in the planning profession (and in urban society more broadly), and discuss how different people have experienced and perceived various urban settings. This knowledge is as important as an understanding of the technical properties and operations of the city.

The identity groups covered in this book do not constitute discrete categories of oppression. On the contrary, they overlap and intersect a fair amount. In fact, they are produced by a series of intersecting systems: patriarchy, capitalism, white supremacy, heteronormativity, imperialism, and so on – which pitch people against one another and undermine the potential for solidarity among the disadvantaged and the oppressed. The adoption of an intersectional lens is another post-modern characteristic of this book.

Intersectionality and plurality – that is, giving equal voice to many groups in a single book – produce obvious contradictions. For example, can wealthier women (of any race) overlook their class interests in order to align with a cross-cutting movement such as feminist planning? Can LGBTQ+ communities and religious minorities that consider homosexuality a sin but have common class interests make good neighbors? Can radically different conceptions of urban land (kin vs property) among Indigenous people and Anglo settlers be reconciled? What if white women's need for safety while walking at night leads to more policing and even incarceration of Black and Brown men?

Besides inter-group conflict, members of the same marginalized group may have vastly divergent experiences in the city. An American-born Latino boy growing up in an U.S. barrio will face different barriers compared to a Latina woman arriving in Australia from Mexico as a university student and living in a dorm. Poverty compounds the mobility and accessibility issues faced by people with disabilities; for example, a blind person with more financial resources may use much more convenient (but costlier) taxi services to move around whereas an impoverished counterpart may have to navigate bus services. Moreover, each group contains a myriad subcategories, which sometimes compete with one another. For example, does the contemporary British city offer more or less opportunity to a darker-skinned Black woman who is native to London and rents a flat compared to a lighter-skinned Black African man who has recently migrated and owns a home?

According to postmodern theoretical paradigms, such as advocacy, transactive, collaborative, and communicative planning, these contradictions should be resolved through political processes involving negotiation, mediation, and "counter-plans." In these processes, planners are expected to act as intermediaries, balancing reason and empathy, so that polarizing debates turn into consensus-seeking dialogues.

A modernist approach?

In other ways, this book's approach is thoroughly modern, as opposed to postmodern, following an academic tradition rooted in the Enlightenment. The book rejects the postmodern view that truth is "inexistant" or "unknowable."[5] While planning outcomes are often determined by political power, discursive frames, and, to put it crudely, money – the classic planning history and theory cannon, which has been gradually formulated by academic planners, is assumed to represent an objective and cohesive metanarrative. The book aims to expand and enrich, rather than subvert, the canon – by recounting in more detail and in a more systematic manner the urban experiences of women, the poor, people of color, LGBTQ+ communities, people with disabilities, older adults, children, religious minorities, immigrants, Indigenous people, and colonized peoples, starting in the late 19th century and following into the present.

A modernist approach to theory does not mean that the book endorses a return to the key modernist paradigm in practice – comprehensive-rational planning. This is clearly impractical for today's rapidly transforming cities. The modernist enthusiasm for dominating and exploiting nature is also undesirable and even disastrous. Neither does the book advocate for a return to dated, modernist urban design, as exemplified in British postwar New Towns or Chandigarh, a regional capital in India. Rather, modernity here refers to applying empiricism and logic in discussing planning history and theory. These features do not preclude care, compassion, and consideration of others – human "others" but also animals, plants, and the natural environment.

A book structured in this manner has been missing until now[6] – but not because academic planners have been unwilling to adopt a critical lens. On the contrary, planning academia has emerged from, and continues to follow, a progressive tradition concerned with social justice, environmental protection, well-being, and even urban beauty. In a sense, planning theory has served as a vanguard for planning practice. The latter has been varyingly progressive and conservative, with great inconsistencies depending on epoch, country, city, or even district. While at certain times and places planning practice has sided with the marginalized and disadvantaged, in many other cases it has been a tool of the systems of power which govern the Anglosphere.

The authors strive to present robust scholarship based on the available historical evidence rather than subjective accounts drawing from their opinions, feelings, or intuition. Reliance on scientific rigor is a key characteristic of modernity. As urban planners and human geographers, the authors are more concerned with the material aspects of urban living which harm certain groups – than with verbal violence or implicit biases in discourse. Phenomena such as gentrification, displacement, housing unaffordability, curtailed mobility and accessibility, environmental dumping, digital surveillance, climate migration, and, in former colonies, informality, are as problematic today as urban renewal, redlining, school segregation, white flight, Indigenous land disposition, and poor sanitation were in earlier eras. These are quite blatant issues rather than debates over subtle or unconscious failings in speech and writing.

Another modernist characteristic is the book's fundamental optimism about the future. Planners, as futurists, have to believe that a city premised on addressing the diverse needs and wants of its entire population is possible – certainly in a wealthy and democratically governed region like the Anglosphere. While a dose of skepticism is healthy (and has been crucial to human Enlightenment and liberalism), post-modern hopelessness and cynicism about the possibility of urban progress is incompatible with the planning profession.

Identity politics vs individuality and universality

The book does not necessarily espouse "identity politics." Each person is unique and the basic premise of planning is that cities need to work for everyone at the individual level. People's collective urban experience does not necessarily trump individuality and universality. Needless to say, there is much ideological and intellectual diversity within identity groups, and not all individuals within a group share the same views on planning issues. Furthermore, divisions, inequalities, and struggles based on class and wealth are central to planning theory; the discipline has historically engaged in a broad critique of capitalism and its knock-on effects. All the authors touch upon problems faced by the poor, and one chapter is entirely devoted to the topic of poverty and homelessness.

Where possible, the book editor invited contributors who identified with the group that they covered. This choice was not made for the purpose of performatively "feminizing" or "colorizing" the curriculum but simply because women and Indigenous people have personal experience in navigating the city in a female or Black body while men and settlers of Anglo descent do not. While all the contributors, as well as the editor, are academics, which by default assigns them privileged status regardless of background, including a variety of identities and voices in a book about planning history and theory helps avoid biases and paint a more complete picture of reality.

The implication is not that only those who are themselves migrants can write about migration issues, only parents can insist on child-friendly cities, and only people with disabilities are allowed to critique disabling urban environments. Trained researchers are capable of transcending their cultural biases or social positioning to understand the perspective and plight of an identity group that is not their own – provided that they are willing to conduct diligent and ethical fieldwork and are, above all, humble, curious, and open-minded. Moreover, all planners (academics and practitioners) have a professional responsibility to advocate for cities that are more just, ecological, accessible, healthy, beautiful, and joyful.

While providing evidence of major discrimination and oppression by the planning systems, the chapters paint a nuanced and changing picture of the urban histories of particular groups. For example, Indigenous people have fared quite differently in settler states – although they have been largely disposed of their lands nearly everywhere. New Zealand's Māori, being more numerous and having better capacity for resistance, have always commanded greater power in the body politic than First Nations Canadians, Native Americans, or Aboriginal Australians. Similarly, the urban experience of women varies a great deal depending on their class and other identity markers that intersect with gender. With regard to religious minorities, Jews were probably the most persecuted and segregated group in the Anglosphere before WWII, whereas now Jewishness is much less relevant, and ghettos are non-existent; instead bigotry is mostly targeting Muslims (of various races), with veiled women physically attacked in supermarkets, beaches, and other public places.

Group identity is fluid rather than fixed – which is another reason why everyone, regardless of group belonging, should have a right to the city. For example, as the chapter authors note the LGBTQ+ label is fairly recent: in the early 20th century, homosexuality was an act rather than an identity operating on a spectrum. Certain psychological disorders or disabilities, such as hysteria (once thought to be common among women), are no longer considered as diagnosable conditions deserving isolation in specialized facilities. Physical disability can be lifelong or temporary; for example, from a transport perspective pregnant women can be categorized as disabled in the sense that they need free seats available on public transport, longer pedestrian crossing signals, and escalators or elevators. A migrant identity is sometimes temporary too, with some people assimilating into, and even coming to identify very strongly with, the hosting Anglo-American culture.

With regard to race, in the U.S. immigrants of Eastern European, Southern European, and even Irish descent were not even considered to be "white" in the first part of the 20th century – in the sense of being treated the same as Americans of Anglo-Saxon ancestry in the job and housing markets. Today, it is unconscionable to accuse a poor industrial worker in a British or American "rust belt" town of having "white privilege" or "male privilege." While the book provides exhaustive evidence of how the planning systems have marginalized and segregated people of color, urban history would be reductionist if it dismissed the struggles (and contributions) of whole swathes of white people.

From another perspective, development planners from the Anglosphere, who offer technical advice and control the funding provided to Sub-Saharan Anglophone Africa and the Indian Subcontinent, are often perceived as privileged here – whether they are Asian women, Black men, queer persons, or otherwise marginalized at home. Although Empire has fallen, Anglo-American wars in the 21st century have undermined urban progress both in the Anglosphere and in newly occupied regions. However, former colonies have their own forms of structuring urban advantages and disadvantages, in addition to skin color, gender, and wealth: ethnicity and tribalism in Sub-Saharan Anglophone Africa and socioreligious caste in the Indian Subcontinent, for example. While colonialism has scarred these places, perhaps forever, planning failures here are not always due to the post-colonial condition but can also be connected to discriminatory and hierarchical structures within the native cultures, which have preceded colonization.

The book points to much unfinished business in planning. Movements such as Black Lives Matter and #MeToo have emerged in reaction to the fact that people from marginalized groups are still being harassed and even murdered in public space. Even as it highlights problems of and in planning, the book is *not* meant to "shame" or "reeducate" planning students, researchers, and practitioners – in the style of authoritarian governments. Nor does the book endorse sweeping generalizations and oversimplifications such as "all men are complicit in the fact that cities are poorly planned from the perspective of women" or "all whites are responsible for the physical segregation or displacement of Black and Brown people." While the chapters may be confronting, the purpose of the book is *not* to plant discord, fuel paranoia, or foment more outrage in a region still reeling from the effects of Brexit, the Alt-Right, and other divisive movements.

Importantly, the book does not sanctify victimhood on the basis of group identity. Painting women, minorities, and former colonial subjects as victims does not dignify them. Besides, as the chapters recount, persons belonging in disadvantaged groups have hardly been passive or self-sabotaging recipients of injustice. With the obvious exception of children, they have organized, resisted, and fought for their right to the city and against racism, classism, sexism, ageism, homophobia, disablism, and imperialism. Rather than an exercise in negativity, the book is an invitation for the planning community to engage in more vigorous and constructive activism on material issues of import to the groups featured here.

Book contributors recognize the undeniable material gains made by various social movements and the positive effects of affirmative action policies and practices in the city. Discrimination – based on race, gender, religion, national origin, disability, and sexual orientation – is now illegal throughout the Anglosphere. Colonialism, child labor, and slavery have been overcome, whereas the class system, female subjugation, racial segregation, and enforced heterosexuality have been challenged in a major way. As the case studies show, examples of successful participatory and inclusive planning processes abound.

From a cultural (rather than legal) perspective, gender roles have changed radically within a few decades. Female planners now comprise a large and growing proportion of the profession, and women are much more visible in public space and in the public sphere. In fact, the need for some urban services and facilities (kindergartens, food deliveries, domestic cleaning agencies, nursing homes) is a direct consequence of women's withdrawal from unpaid reproductive work – housework, cooking, childcare, elderly care, and so on. And upward mobility has been increasingly common among Black, Latinx, and Asian urbanites in the past 50 years – thus changing the face of many urban neighborhoods. These gains are not a mirage. It is important to celebrate and sustain them.

Book description

Chapters come from all parts of the Anglosphere as well as two formerly colonized regions in which the use of the English language is widespread (the Indian Subcontinent and Sub-Saharan Anglophone Africa). To allow for easy cross-comparison, chapters follow a similar chronological structure (set forth in a table at the end of this chapter). The authors provide insights into the core planning issues in each time period, and review the different stances and critiques. Chapters synthesize and interpret the existing literature rather than providing new empirical material. In so doing, planning history and theory are treated as inseparable.

To assist planning students and instructors, each chapter includes the following pedagogical features: (1) a boxed case study which presents a recent example of positive change to showcase theory in practice; (2) a table which lays out an alternative planning history and theory timeline for the group covered in the chapter; and (3) suggestions for further study comprising non-academic sources such as books, websites, and films.

A short preview of each chapter, provided by the respective authors, follows.

Chapter 1, by *Sewell*, posits that women's life experiences have been greatly affected by the assumptions that (largely male) planners have made. Most notably (a) the belief that domestic life and waged work life are and should be separate, (b) the general discounting of feminized labor, and (c) the use of the male-headed nuclear family as a norm. Women have pushed back, claiming public space as shoppers, workers, and citizens, and have fought to make cities more sanitary, beautiful, convenient, and safe. In the late 1800s, women were increasingly active participants in the public sphere and public space, fighting for the vote, founding reform organizations, shopping downtown, and going to work in shops and offices. In the 20th century, women expanded their roles as voters and workers, but found the interconnected complexity of their domestic and working lives ignored by planning which used zoning in ways that made it difficult to balance work and home, a move that culminated in the growth of the suburbs in the post-World War II era. Beginning in the 1960s, a new wave of feminism pushed back against sexist assumptions, and eventually adopted an intersectional lens, while the dogma of planning was being challenged more broadly. However, even as planning has begun to embrace complexity and participation, women's needs and voices have remained marginalized.

In Chapter 2, *Howard* notes that American urban planning has wrestled with the presence of the poor and homeless in various ways since the birth of the field. Issues of crowding and sanitation in the 19th century prompted the launch of urban planning as a profession. During the early 20th century, the crisis of the Great Depression prompted reconsideration of the role of the government in the nation's housing market. As a result, the postwar years witnessed the birth of public housing and the maintenance of skid row districts, while urban renewal cleared the way for more monied classes. During the 1960s and 1970s, the welfare state expanded to accommodate more families and individuals. The neoliberal policies of the late 20th century and early 21st century, however, fostered public-private growth coalitions that displaced many of the poor. In recent years, gentrification of desirable urban centers has left many Americans homeless and others priced out of their former neighborhoods.

In Chapter 3, *Muller Myrdahl* provides much needed nuance to the story of sexual- and gender-diverse communities (as described by LGBTQ+, for lesbian, gay, bisexual, transgender, queer, plus) in North American cities. This story is often told as a simple trajectory: what was once hidden is now visible, and visibility has enabled inclusion. However, simple narratives gloss over complex histories and contemporary realities. Visibility is complicated by myriad other factors like race, class, and community cohesion. This chapter highlights key concerns in each chronological period, from the emergence of sexual identity to concerns over the demise of the gay village. Today, LGBTQ+ inclusion remains a contentious element of the municipal agenda.

Chapter 4 by *Johnson*, *Tucker*, and *Watchorn*, deals with issues of disability in the Australian city. If planning involves the regulation of people and activities in space as well as service delivery, then the definition and management of people with disabilities has intersected at critical points with these systems. From early colonization in Australia, those unable to effectively participate in work were confined to their homes, streets, or prisons. The first planned interventions created large institutions in isolated grounds where those defined as deficient were confined. The regulation of the "disabled" reached its nadir with the eugenics movement of the early 20th century. As planning began improving the lives of urban citizens, the effective incarceration of those with intellectual disability continued. Those with physical impairments were gaining freedoms though were still profoundly limited in their participation in planning and society. With a series of United Nations Declarations and the emergence of disability advocates, the rights of people with disabilities were increasingly asserted, leading to deinstitutionalization and more autonomy in accessing services, culminating in the creation of Australia's National Disability Insurance Scheme. Despite this revolution in the provision of services, inaccessibility in the built environment remains, along with the spatial segregation of many. There is still much to do by planning to ensure access and inclusion within the city.

Chapter 5 by *Lewis* explores the origins and changing nature of ageism in North America. Planning for the needs of older adults in cities has been part of the vocabulary of planning research and practice for well over a decade. The coronavirus pandemic has only too recently and tragically revealed the extent to which our most vulnerable citizens have been marginalized by policy makers and by society more broadly. The alienation of those who have come to be considered the least productive members of society is rooted in several factors which include demographic changes in the 19th century, the advent of Positivist professions that have reduced the ageing experience to a "problem" to be solved through scientific intervention, and the emergence of a form of compassionate ageism that has framed older adults as both deserving and needing segregation from the wider community. The needs of older adults have been largely peripheral to the scholarship and practice of planning which, for much of its history, has been either the purveyor of built form for young, working class families or, more recently, rational advisor to other professions on how to plan for the aged. However, planning practice and scholarship could contribute more directly to the articulation of plans or visions that define what it means to "age in community" and challenge the segregation of older adults that has become too prevalent in recent history.

In Chapter 6, *Krishnamurthy*, *Wood*, *Strachan*, and *Peacock* discuss the needs of children in the British city. Planned environments play a central role in children's lives, by creating opportunities and constraints for physical activity, mobility and positive health outcomes. Urban planning can facilitate the creation of functional and inclusive environments for all young citizens. However, despite burgeoning global interest in children's well-being in cities, there remains little clarity or institutional support for children's needs or active role in urban planning processes.

This chapter examines the history of degrees of separation between the idea of childhood (as a social construct) and developments within urban planning in the United Kingdom and beyond. By drawing links between the two, the chapter shows that planning can assume a leading role in facilitating children's freedom and independent mobility, addressing concerns around health, inequalities, human rights and safety. This connects to increasing awareness around children's rights, engagement processes incorporating children's views, new methods of working with children, and the global acknowledgement of the multiple crises of childhood.

Chapter 7 by *Dunn, Itaoui,* and *Ngui* documents a consistent theme of religious inequality in Western cities, including cities in Australia. This is demonstrated clearly for non-Christian religious groups, who have struggled to establish their places of worship. For Muslims, in cities like Sydney, the effect of Islamophobia has been stark. Planning for non-Christians in Western cities has faced prohibition, attack, and frustration. The legacies of Christian supremacy have persisted through secularization, disestablishmentarianism and multiculturalism. British colonization and white supremacy are implicated in this persistence. The religious landscape remains dominated by Christian infrastructure. On the positive side, the authors note that new planning scholarship, influenced by the postmodern paradigm, has provided a conceptual path toward substantive equality and inclusive citizenship in religiously diverse societies. They also provide practical recommendations for addressing geographies of exclusion among religious minorities, while also highlighting the need to strengthen alternative planning practice such that religious plurality and diversity is freely expressed and celebrated.

In Chapter 8, *Heil, Turner,* and *Wilson* highlight a contradiction in the discipline of the U.S. urban planning, which appears when viewed through Black urban history and politics. A field that purports to improve cities for people's well-being has, throughout its history, produced interventions into urban life that harm Black people and communities. This chapter examines the relationship between urban planning and people of color in the United States. The authors interrogate: (1) how practices of planning and city-making have contributed to racialized disparities in wealth, health, and opportunity and (2) how Black people have organized communities to build lives and spaces in the city amid discriminatory policies. Beginning with the late 19th and early 20th century, the chapter documents the active role of urban planning in producing the segregated city. Moving to the postwar era, the authors examine urban renewal and the destruction of racialized neighborhoods. They then chronicle how planning in the 1960s and 1970s, engaged the demands of Black liberation movements. Finally, moving to the end of the 20th century, the story turns to the "urban crisis" era as neoliberal policy left Black communities under-resourced and exposed to environmental discrimination. The chapter concludes with a discussion of the rise of gentrification and current debates over the role of planning in supporting racial justice.

Chapter 9 by *Huerta, Buelna,* and *Buelna* addresses the history and practice of anti-Mexicanism in the U.S. cities. While racism continues to be viewed as a white-Black paradigm, it is imperative to shine light on the case of people of Mexican origin, who have historically been racialized, marginalized and abused by the dominant culture and its repressive agents and institutions (e.g., courts, police forces, immigration agency, government). While the American Southwest once belonged to Mexico, Mexican immigrants continue to be demonized and stereotyped as "illegal aliens" and "invaders." Even those individuals of Mexican origin that have acquired the U.S. citizenship are reduced to being "foreigners in their own land." Despite their enormous economic, cultural, artistic and social contributions to the U.S. cities, Mexican Americans are regarded as second-class citizens. By focusing on the urban experience of Mexican immigrants (and Mexican Americans or Chicana/os), the chapter shows how planning has failed to challenge systemic racism in general and anti-Mexicanism in particular.

In Chapter 10, *Parsons* points out that planning was and still is a critical actor and principal structure in the colonization and dispossession of Māori in Aotearoa New Zealand. In this chapter, one tribal group's (Ngāti Whātau Ōrākei) experiences in engaging with the settler-colonial government and its planning regime in the city of Auckland (Tāmaki Makarau) is traced from its origins in the 19th century through the present. The chapter begins by chartering how Ngāti Whātua initially supported the establishment of the city and the settler-government's policies. However, as the tribe was dispossessed of their lands, they increasingly challenged and resisted colonial intrusions and demanded that their sovereignty and agency be recognized and protected. Next, the chapter provides insights into how the growth and maturation of Auckland as a city involved the introduction of government policies and planning practices that aimed to culturally assimilate Māori into white (European) culture. Then, the chapter provides insights into mounting Māori protests against further acts of dispossession and the re-emergence of Māori rights following legal settlements with the New Zealand Government. Lastly, the chapter reviews the opportunities and challenges that Māori face in seeking to exercise their knowledge, values, and sovereignty within 21st century planning practices.

Chapter 11, by *Ghertner* and *Banerji*, discusses the trajectories of urban planning policies in the Indian subcontinent from the British colonial period into the early 21st century. The chapter suggests that informality, often seen as a post-liberalization feature of privatized and splintered planning systems, has been a defining feature of planning the Indian city for nearly two centuries. By analyzing the persistent conflicts between the technical and the political, originating as early as the mid-19th century in disputes over municipal governance, the chapter suggests that an alternative planning discourse has silently been operating in the Indian subcontinent. It operates to persistently challenge the normative underpinnings and centralizing impulses of planning principles, inserting subversive demands of the urban majority into the technical operation of planning practice. Urban planning, on this basis, is not something unevenly or fitfully applied in the post-colony, as is often suggested, but as an alternative terrain of democracy in action. India's presumed planning "failures" require careful study both for what they do not achieve and for their effectiveness in prompting political experiments that unevenly incorporate the urban majority.

Finally, Chapter 12 by *Myers* and *Owusu* examines urban planning dynamics in three African countries where English is the official language of planning (Ghana, Kenya, and Tanzania). Despite great variation and diversity, these settings exemplify many themes common to planning in the former British colonies of Africa south of the Sahara. The struggle to assert an African urban identity for planning faced tremendous challenges after independence, including overlapping economic crises in the public sector and global forces for privatization. Many government-led planning efforts have been unsuccessful, and instead many cities have grown in largely informal ways. Even with a 21st century resurgence of state planning efforts and infusion of foreign capital, the unplanned informal aspects of urban development remain strong. The authors argue that it is often in the informal zones of cities such as Accra, Nairobi, or Dar es Salaam that alternative planning finds roots.

Conclusion

Contemporary cities are marked by multiple crises, which cannot be fixed through incremental planning. But radical activism – in planning as in other arenas – cannot occur without solidarity and coalitions that recognize and highlight our shared humanity. Feminist planners must ally not only with migrants and Indigenous people, disability and LGBTQ+ activists, people of color, anti-imperialists, children's rights organizations, aged care advocates, and religious minorities, but also with labor unions and environmentalist groups to begin to envision sweeping solutions that transform currently oppressive states, societies, and economies in conjunction with tangible urban spaces.

Period	Planning history and theory canon in the Anglosphere (and some former colonies)
Late 1800s–1900	*Birth of planning*
	Late-stage colonialism, rapid industrialization, sanitation and water supply, building and fire regulations, tenement housing, social reforms, first urban plans, urban park movement
1900–1945	*Formalization of planning*
	Some colonies gain independence, nation building efforts, modernism, utopian urban models (Garden City, City Beautiful, etc.), Great Depression, exclusionary zoning, first planning programs at university level
1945–1965	*Growth of planning*
	Postwar reconstruction, disintegration of British Empire, economic growth, Keynesianism, social welfare, suburban expansion, New Towns, motorization, urban highway expansion, shopping malls, consumerism, first metropolitan/regional planning schemes, rational-comprehensive planning, technocratic planning, modernism, incrementalism, mixed scanning

(Continued)

Period	Planning history and theory canon in the Anglosphere (and some former colonies)
1965–1980	*Midlife crisis of planning* Environmentalism, economic crisis, oil crisis, international and domestic migration, feminism, Civil Rights, Indigenous people's rights, disability rights, gay and lesbian rights, children's rights, advocacy/equity planning, radical planning, transactive planning, post-modernism
1980–2000	*Maturation of planning* Globalization, de-industrialization, service economy, neoliberalism, austerity, privatization, public-private partnerships, growing social inequality, gentrification, pursuit of "sustainability," new planning models (New Urbanism, Smart Growth, TOD, etc.), strategic planning, communicative planning, collaborative planning
2000–present	*New planning crisis* Climate breakdown, populism, mass migration, hyper-gentrification, rentier urban economies, urban boosterism, extreme consumerism, housing unaffordability, new technologies, digital platforms, Smart Cities, high inequality, the Just City concept, Great Recession, UN Sustainable Development Goals, COVID-19 pandemic

Source: Based on Fainstein and Campbell (2011) and Johnson (2017).

Notes

1 The term Anglosphere refers to a group of five highly developed, English-speaking nations (Australia, New Zealand, the United Kingdom, the United States, and Canada), which share cultural and historical ties to England, and continue to maintain close co-operation. Sometimes, the Republic of Ireland is included in the Anglosphere as well.

2 It is beyond the scope of this Introduction to unpack all the theories, models, landmark moments, and concepts listed in Table 1. Interested readers can consult other texts (for example, Faludi 1978; Mandelbaum et al. 1996; Taylor 1998; Brooks 2002; Archibugi 2007; Healey and Hillier 2010; Fainstein and Campbell 2011; Hardin and Blokland 2014; Hall 2014; Fainstein and DeFilippis 2015; Hein 2017; Burchell and Sternlieb 2017; Allmendinger 2017; Jayne and Ward 2017; Haughton and White 2019; Beauregard 2020; LeGates and Stout 2020).

3 A thorough discussion of postmodernity and its manifestations in the built environment is provided by Harvey (1989).

4 Based on the definition of the movement provided by the Encyclopedia Britannica (www.britannica.com/topic/postmodernism-philosophy).

5 For a critical comparison of modernism and postmodernism and their practical applications, see Pluckrose and Lindsay (2020).

6 An exception is a volume edited by Sandercock (1998) on multiculturalism.

References

Allmendinger, P. 2017. *Planning Theory* (3rd edition). London: Bloomsbury.

Archibugi, F. 2007. *Planning Theory: From the Political Debate to the Methodological Reconstruction.* Cham: Springer.

Beauregard, R. 2020. *Advanced Introduction to Planning Theory*. London: Edward Elgar Publishing.

Brooks, M. 2002. *Planning Theory for Practitioners*. New York: Routledge.

Burchell, R., Sternlieb, G. (eds.) 2017. *Planning Theory: A Search for Future Directions*. New York: Routledge.

Fainstein, S., Campbell, S. (eds.). 2011. *Readings in Planning Theory* (3rd edition). New York: Wiley-Blackwell.

Fainstein, S., DeFilippis, J. (eds.). 2015. *Readings in Planning Theory* (4th edition). New York: Wiley-Blackwell.

Faludi, A. (ed.) 1978. *Essays on Planning Theory and Education*. Oxford, UK: Pergamon Press.

Hall, P. 2014. *Cities of Tomorrow: An Intellectual History of Urban Planning and Design Since 1880* (4th edition). Cambridge, MA: Blackwell.

Hardin, A., Blokland, T. 2014. *Urban Theory*. Los Angeles: Sage.

Harvey, D. 1989. *The Condition of Postmodernity*. Cambridge, MA: Blackwell.

Haughton, G., White, I. 2019. *Why Plan? Theory for Practitioners*. London: Lund Humphries.

Healey, P., Hillier, J. (eds.) 2010. *The Ashgate Research Companion to Planning Theory*. Surrey, UK: Ashgate.

Hein, C. (ed.) 2017. *The Routledge Handbook of Planning History*. London: Routledge.

Jayne, M., Ward, K. (eds.) 2017. *Urban Theory: New Critical Perspectives*. London: Routledge.

Johnson, L.C. 2017. Australian planning texts and Indigenous absence. In S. Jackson, L. Porter, L.C. Johnson, Planning in Indigenous Australia: from imperial foundations to postcolonial futures, pp. 34–51. London: Routledge.

LeGates, R., Stout, F. (eds.) 2020. *The City Reader* (7th edition). New York: Routledge.

Mandelbaum, S., Mazza, L., Burchell, R. (eds.) 1996. *Explorations in Planning Theory*. Livingston, NJ: Transaction Publishers.

Pluckrose, H., Lindsay, J. 2020. *Cynical Theories*. Durham, NC: Pitchstone Publishing.

Sandercock, L. (ed.) 1998. *Making the Invisible Visible: A Multicultural Planning History*. Berkeley, CA: UC Press.

Taylor, N. 1998. *Urban Planning Theory Since 1945*. Los Angeles: Sage.

1

WOMEN

Complex lives in the patriarchal city

Jessica Ellen Sewell

Introduction

Women have been part of cities since their beginning, but they have been marginalized by the decisions that planners have made. The urban experience has provided women with pleasure and freedom, and urban public spaces have been a necessary respite from male-dominated households and workplaces, in spite of the fact that cities have not been planned to accommodate women. Nineteenth-century middle-class ideals that pinned white women to domestic space, ideally in class- and race-segregated neighborhoods or the suburbs, have shaped how women's roles in cities have been imagined and shaped well into the 20th century. Planners sought to make the city more orderly and less dangerous, often seeing women, particularly independent women, as a source of disorder that needed to be controlled and contained (Wilson 1991). As male planners sought to make order out of the disorder of crowded industrial cities, they began by assuming the nuclear family and the male wage earner, separating male spaces of work and female spaces of domesticity. Starting from their own experiences, they planned public parks that prioritized boys' play, public transportation that prioritized men's commutes, and public spaces that were a poor fit for the needs of women and of families. Women have pushed back against these constraints, making a strong argument for their right to the city and pointing out the changes that must happen both to cities and to the tools of urban planning to make that claim a reality.

Late 1800s–1915: women claim the city

The mid-19th century was the heyday of the ideology of separate spheres, in which women and men were imagined as distinctly different creatures, with men associated with the public sphere of work, politics, public space, and the city,

DOI: 10.4324/9781003157588-2

and women associated with the private sphere of domesticity, children, and the home. During this period, elites moved to separate the spaces of work and home, living in increasingly purely residential neighborhoods within the city or moving to homes in elite suburbs. While this ideology never fit the majority of lives, its limitations became ever more visible in the later part of the 19th century with the growth of mass production and associated mass consumption. Cities were transformed with new and growing downtown shopping districts, anchored by department stores, a business type invented in the 1850s in France and common throughout the Global North by the 1890s. By the late-19th century, the life of elite white women, even if focused on home and domesticity, necessitated regular visits to the heart of the city for shopping, club meetings, and other forms of consumption, leisure, and service. The businesses and institutions they patronized often created separate spaces for women, including women's windows at banks and post offices, women's waiting rooms in train stations, women's reading rooms in libraries, women's matinees in theaters, and even special women's cars on public transportation. These gender-segregated spaces were designed to protect white women from the dangers of sharing urban space with men, containing them in feminine bubbles, but most of them quickly became outdated as women became seen as natural denizens of the downtown shopping district and related public spaces. In these new urban spaces of consumption, elite and middle-class women were often served by working women, whose presence was essential to feminizing the department stores, tea rooms, and other public spaces that catered to elite women. The result of the growth of downtown consumption was an expanded presence of women in public, shopping, working, eating, taking public transportation, and walking the streets.

This period also saw an increase of women working in offices and factories, supporting the increased bureaucracy and production of the modern industrial era. Working-class waged employment replaced the household-based production work women had done before the Industrial Revolution with paid labor as domestic servants, shop and restaurant workers, piece workers, factory laborers, clerical workers, and sex workers (Wood 2005). While elite women were white and mostly non-immigrant, working women were more diverse, including immigrants and, in the United States, Black women who were fleeing the violence and deprivation of the American south. Many women workers were young women who came to cities from rural areas and small towns and often lived in boarding and lodging houses, usually managed by women. They existed outside of the bounds of the nuclear family that was the core of women's imagined sphere, and typically in working-class neighborhoods that housed both respectable and illicit businesses (Meyerowitz 1991; Deutsch 2000). Young working women claimed space in the city as consumers in their own right, frequenting cafés, dance halls, theatres, and shops, as well as public transportation and the streets (Peiss 1986; Meyerowitz 1991).

In spite of women's presence in the public spaces of the city, from department stores to local main streets to public transportation and their increasing participation in the waged workforce, the ideology of separate spheres continued to influence how experts, including urban planners, thought about women's role in the city. Women in public were often treated as matter out of place; the term "public woman" was historically a euphemism for prostitute. While women's presence as shoppers and as workers helped to ameliorate the assumption that a woman in public was immoral, women did not have the freedom to simply enjoy life in public. Unlike male flâneurs, they had to be purposeful and careful about where in the city they appeared and how they behaved in public (Sewell 2011). As most public space was considered male by decision makers, amenities such as public bathrooms were planned to accommodate men only. The lack of provision of restrooms for women in public space curtailed their ability to spend extended periods of time there and signaled to them that they were not welcome, unlike men who were provided with public urinals and, in addition, could more easily urinate in public places (Flanagan 2018).

In the late-19th and early-20th century middle-class and elite women also made use of their moral authority as women and as mothers to reshape aspects of the city and their role in it through settlement houses, philanthropic organizations, improvement clubs, and political campaigns. Settlement houses, which began in England in the 1880s and were quickly adopted in the United States, were located in working-class neighborhoods and served as a location for services including childcare, education, physical education, and meeting rooms. They were also a base for live-in women reformers to serve and study needy, often immigrant, populations. Along with other philanthropic institutions, settlement houses provided an opportunity for educated women to claim expertise as teachers and social workers; to influence the lives of the communities they served; and to push for changes in policy and practices surrounding wages, child labor, housing quality, and the public provision of services such as bath houses and playgrounds.

In addition, women in philanthropic institutions created new spaces for women and children in the city by building settlement houses, orphanages, kindergartens, and playgrounds (Spain 2001; Gutman 2014). These institutions actively made space for women and children in the city, at times taking over masculine space, as in the West Oakland Free Kindergarten in California, which took over a building that had been a saloon facing the entrance to the railroad yards (Gutman 2014). The Young Women's Christian Association (YWCA) focused particularly on the creation of spaces that served working women, providing safe and affordable housing for young working women in grand purpose-built urban edifices. These were built in over 200 cities, both large and small, across the United States as well as internationally, and served Black and Chinese communities as well as white working-class women. In addition to housing, YWCAs provided vocational training to young women, particularly in the domestic arts, and provided space for women's clubs, parlors and reading rooms, recreation facilities, and day care services, all serving women (Spain 2001). These women-run spaces provided public and semi-public space for independent single working women who were not safely provided for in cities based on the idea that respectable women should not be independent. Through their service, settlement women and other charitable workers both claimed a public role for themselves and also made changes in the city on behalf of the women and others they served.

Elite women also strove to make a mark on cities through civic-minded women's clubs which sponsored urban beautification campaigns, founded public institutions, and pushed for political reforms, especially those related to women and children. In the United States, many of these clubs were allied with the General Federation of Women's Clubs (GFWC). This association, founded in 1890, coordinated women's clubs in order to strengthen women's civic influence. The GFWC helped elite club women to shape cities physically and socially through the creation of institutions such as libraries, art museums, and schools (Enstam 1998). Women's clubs were also actively engaged in municipal housekeeping, pushing to make cities cleaner and more orderly. They argued for clean food and water and improved sanitation, but also attempted to clean cities of vice and crime in movements against prostitution and alcohol. In so doing, they brought their own white, middle-class ideas about appropriateness to bear on the lives of those poorer than they, often exacerbating rather than solving the problems poorer women faced.

Women used their role as civic housekeepers and as active participants in the public life and space of cities to make a claim for the right to full citizenship in women suffrage movements in the English-speaking world during this period. Suffragists claimed urban space for political speech, setting up downtown headquarters, holding public meetings, and speaking in the streets and parks (Sewell 2011).

1915–1945: modern working women

In the period encompassing the two world wars and the interwar period, women more fully entered the public realm as workers, as shapers of reforms and institutions, and (in the United States and United Kingdom) as newly enfranchised citizens. At the same time, new planning ideas and tools, including zoning and comprehensive planning, gained more power to enforce a worldview in which women were associated with the domestic sphere.

Both World War I and World War II had significant consequences for women in the workplace. With many men leaving their positions to serve in the military, women entered the workforce in increasing numbers, both to fill vacancies and to fill new clerical positions created by increasing bureaucracy. The 1910s saw an increase in particular of women in professional positions, with white women newly taking on managerial positions and such previously all-male jobs as ticket agents and trolley car conductors. As white women moved up the career ladder, in addition to taking on previously masculine positions as laborers in factories, an increasing number of women of color in the United States quit working in domestic positions and moved into better paying and more stable positions as laborers in factories, as well as in clerical and service positions in Black-owned and Black-serving businesses (Enstam 1998). The economic boom after World War I meant that, rather than being pushed out of the labor market when men returned from fighting, more women were pulled into waged work, and during the Great Depression, women, who were largely in feminized positions, did not disproportionately lose their positions. The labor demands of the wartime economy in World War II further pulled women into waged work, with the number of women working for wages in the United States increasing by 43% between 1940 and 1945 (DuBois & Dumenil 2009).

As they moved into managerial positions and were increasingly trained in the professions, women joined newly created women's professional associations, while women workers in feminized fields such as waitressing, garment making, and teaching claimed authority through unions. Women also gained equal rights as voters in the United States in 1920 and the United Kingdom in 1928, giving them both new political power and new symbolic power as citizens. The interwar years saw women making new claims to modernity and autonomy, but doing so in the context of significantly lower wages for women and a social structure still largely based on the assumption that women's true calling was marriage and motherhood.

The period between 1915 and 1945 saw a significant growth of planning, with the introduction of comprehensive zoning in the United States in 1916, the growth of comprehensive planning, new mechanisms to respond to housing needs, and large-scale economic planning in response to the Great Depression. However, these planning mechanisms were dominated by male planners and built around assumptions that often ran counter to women's real needs. They used as a model a single-family household with a male wage earner; for example, Depression-era policies in the United States actively encouraged laying off female workers if their male family members also worked for the government (DuBois & Dumenil 2009).

Both housing reforms and zoning made life more difficult for women trying to balance waged work, household errands, and reproductive work in the home. In the 1930s, policies in both the United States and United Kingdom encouraged the demolition of substandard tenement housing and moving residents to more purely residential neighborhoods, usually at the edges of cities, which separated women from their social support systems and their sources of casual work (Roberts 1991; Flanagan 2018). In the United States, zoning, which was declared constitutional in 1926, similarly sought to sort and separate different sorts of buildings and uses in the city, using an understanding of the activities of the city based on middle-class and elite male experiences in which home and work are separate. Zoning protected both residential areas and high-rent offices and shops from noxious factories, and above all supported real estate values by creating predictability, but in the process it helped to solidify existing separations of space by class, race, and gender. In the United States, redlining, a federal policy by which loans for real estate in Black, mixed-race, and immigrant neighborhoods could not get federal insurance, combined with zoning to make life particularly difficult for Black women and women of color, who were forced to live in substandard and overcrowded housing with poor transportation links to workplaces. In addition, while the issues of childcare and cooking were addressed in a few projects for women working on the home front during World War II, balancing the competing demands of family and work was largely left to individual women.

1945–1965: redomestication

The postwar era saw a refocusing of planning priorities onto the nuclear family and large-scale systematic approaches to issues of urban growth. While women continued to expand their participation in the workforce, higher education, and civic life, the ideology of feminine domesticity took on a renewed fervor, supported by government policies. Popular culture was dominated by images of familial domesticity and the sterility of the working world for women, while new psychological theories argued that conforming to gendered roles was the essence of mental health. Urban policies supported an accelerated separation between feminized residential districts, particularly in newly built suburbs of single-family homes, and urban centers. Modern planning ideas emphasizing separation of uses which had been articulated during the Depression and World War II gained popularity in the rush to catch up with the pressing needs of growing urban areas.

Modern planning ideals, as expressed in the Athens Charter,[1] railed against the density and haphazard nature of residential areas in cities, and in general the disorderly nature of cities, which they saw as both irrational and unhealthy. They sought to reorder cities by sorting and separating different functions. The aim was to serve what they saw as the three functions of cities: dwelling, working, and recreation, each in its own designated space. The lack of attention to spaces of consumption or to the connections between these places for anyone but a male head of household insured that cities planned by these principles poorly served both women within nuclear families and single women.

The fear of density led to the relegation of residential space to dispersed suburban and exurban locations, whether new towns and council housing in the United Kingdom or all-white suburban subdivisions in the United States. Incentives were provided to demolish the poorer and mixed-use areas of cities that were anathema to modern planners and replace them with carefully sorted spaces, including new highways that cut through lower income areas and connected new suburbs to jobs in the center city. This slum clearance put further pressure on women who were not part of nuclear families, as single-room-occupancy hotels and other spaces that served single working women were targeted for demolition as unhealthy and immoral.

The postwar era saw a new wave of responses to overcrowding and demand for housing. In the United Kingdom, the 1946 New Towns Act facilitated the creation of 32 planned new towns to counteract the growth and density of older cities and to counteract unplanned sprawl. These new towns were planned around the needs of a male head of a nuclear family household, with little attention or respect given to spaces of consumption, which were seen as frivolous rather than necessary, or to women as workers. Decisions about shops were usually focused not on the needs of women as shoppers, but rather on the regulation of shop design and the desires of retailers, leading, for example, to shopping centers with large numbers of steps and no seating areas, which were difficult to navigate for women with children and shopping carts (Greed 1994).

Large new housing estates were also created on the fringes of existing cities, similarly isolating women from work opportunities and enforcing the norm of the male-headed nuclear family. Regulations against running a business out of council flats further hamstrung women who wished to balance running a household and making money. Similar regulations constrained women in the smaller number of public housing projects that were built in the United States in this same period, while women in privately built subdivisions faced covenants that tightly constrained their access to home-based work as well.

In the United States, government policies favored urban renewal, some public housing, and primarily the creation of suburban subdivisions of single-family houses. Suburban developments, with their covenants against any actions that would decrease real estate value, including selling or renting to non-whites, were the preferred investment for federally insured loans, both to developers and to homebuyers. Suburban developments were not only purely residential and white, but were sorted by income level and life stage as well, further isolating the women who lived there. In addition, women without children at home, including single women, widows, and women in childless couples, as well as the majority of non-white women, for whom suburban single-family home life was not accessible, were not well accommodated by this housing model that assumed that "woman" was equivalent to "stay-at-home mother." Those who had no access to suburbia because of racial covenants or finances lived in mixed urban neighborhoods, often in rental housing.

The growth of suburbia also tended to alienate white middle-class women from the central city, as downtown shopping and services were replaced by privatized suburban shopping centers. For many women who lived there, the child-focused life of both American suburbia and UK new towns was experienced as a trap, articulated by Betty Friedan in her description of "the problem that has no name" (Friedan 1963; Wilson 1991). At the same time, planners' desire to sort and rebuild the city was being forcefully challenged by Jane Jacobs in *The Death and Life of Great American Cities* (1961). While Jacobs did not take an explicitly feminist point of view, and did not address the ways that the "eyes on the street" she encouraged can function to police women particularly, the diverse and lively mixture she promoted provides a potentially richer and more supportive experience for a variety of women than was provided by suburbia and specialized urban housing developments.

1965–1980: spaces of women's liberation

The late 1960s was a period of questioning and social revolution in many areas. It was an era of antiwar activism; Black, Chicano, Gay, and Disability power; environmental activism; sexual revolution; and women's liberation. Social truisms, including both the modernist ideals of the planning and architecture establishments and the social ideals of marriage, homeownership, and the nuclear family, were brought into question. However, cities continued to transform in ways remarkably continuous with the earlier postwar era, with large-scale urban renewal projects and the construction of public housing through the early 1970s in the United States; the building of highways through and around urban areas; and increasing suburbanization paired with a continued abandonment of cities by both the white middle class and their employers, who moved many businesses to new suburban office parks and other highway-focused locations.

Many women gained a new psychological and legal sense of themselves as individuals, rather than as a member of a family. The advent of the birth control pill, which allowed women to control their own fertility, combined with legal rulings and acts that made birth control and abortion legal in both the United States and the United Kingdom gave women significant new freedom to control their life course and to avoid motherhood if they so desired. An ever-increasing number of women, both married and single, were in the workforce, and their rising educational status meant that these positions were often professional positions that women planned to hold throughout their lives. Women also gained new legal rights as individuals, with US women gaining the right to have any sort of credit in their own name in 1974 and Irish women able to own their homes outright in 1976.

In the context of a new feminist awakening, facilitated through consciousness-raising groups in which women critically examined how sexism and patriarchal ideals affected their lives, women made new claims on the space of the city. Through lawsuits, protests, and marches women asserted their rights to a range of public spaces that had previously excluded them, including public baseball fields, playgrounds, restaurants, bars, country clubs, and universities. Women also laid claim to the street in anti-violence "Take Back the Night" marches beginning in 1978 in the United States and eventually becoming an international phenomenon. In these marches women claimed a right to safety from sexual violence both in the public spaces they marched through and in the private space of the home.

At the same time as women fought for full rights to spaces including streets, restaurants, and previously all-male universities, women also created a range of new spaces to serve them as an alternative or respite from the patriarchal world. Bookstores, coffee houses, and women's centers provided female-centered community space, often in more neglected areas of cities. While these could be seen as a return to the gender-segregated spaces of the 19th-century city, they were run by women for women, and existed in addition to mixed-gender spaces that were accessible to all. Women's shelters and women's health centers not only served women, but also lived the slogan "the personal is political" by bringing issues central to women's lives, such as partner abuse and birth control, out of the household and into new visible public-serving spaces (Enke 2007; Spain 2016).

The 1960s and 1970s saw a significant transformation in women's access to spaces and services and the expansion of privatized services, such as child-care centers and take-out restaurants, that helped make it possible for working women to manage the second shift of taking care of the household (Spain 2016). Through grassroots activism, women made new claims for themselves as urban citizens, opening new doors for themselves as professionals and decision makers, even as they remained at the fringe politically and in the minds of planners.

1980–2000: feminizing the city

The second-wave feminism of the 1960s and 1970s challenged the ways that women were excluded from urban spaces and decision-making positions. This laid the groundwork for women to take a more equal place in society, including in the planning profession. While earlier feminists had created alternative structures and spaces to gain knowledge and power, in the late-20th century women increasingly entered into existing structures of power with the hope of changing them from within. In the urban planning profession, while women were still a minority of planners, they constituted an international network which worked within professional organizations to promote both women in the profession and women's issues in cities. Throughout the fields associated with planning and city-making, there was a belief that feminizing the professions would lead to a significant change in decision-making. However, for the most part the planning profession tended to treat planning for women as a minority issue, an accommodation of specialized needs rather than as an impetus for rethinking norms (Greed 1994). The types of needs most often catered to, by women planners as well as the profession as a whole, were the need for safety in urban space and the needs of women as mothers. These evidence the continuing power of traditional conceptions of white womanhood as endangered and as defined through maternity and family.

The end of the 20th century also saw the continuation and expansion of women's political protests in the context of the conservative political regimes of Ronald Reagan in the United States and Margaret Thatcher in the United Kingdom. Local protests, like the "Take Back the Night" rallies that began in the late 1970s, became an international phenomenon. The fight to maintain abortion rights in the United States was the impetus for several massive marches on Washington DC, as well as local activism defending clinics. In the United Kingdom, women mobilized in a female-only protest against a nuclear missile base at Greenham Common, setting up an encampment there from 1981 to 2000. These protests demonstrate the willingness of women to claim space and voice, specifically as women, to insist for the right to be treated as equal citizens. They also laid the groundwork for further activism, including "Slut Walks" beginning in 2011 and the Women's Marches worldwide in the wake of US President Donald Trump's inauguration in 2016.

The 1980s and 1990s saw an explosion of scholarship that addressed the history and present of women in the city, epitomized by *Women and the American City*, first published as a special issue of *Signs* in 1980, and the Matrix collective's *Making Space: Women and the Man-Made Environment*, published in 1984. This new scholarship critically addressed the historical roots of the patriarchal structure of urban space and decision-making, explored the current status of women in cities, uncovered a history of women shaping and living in cities, and laid out suggestions for remaking both the built spaces of the city and urban planning practice.

The expansion of feminist critiques and women's activism, from the workplace to the streets to the pages of books, took place in the context of a shift in cities from the government-led redevelopment of the postwar era to market-led gentrification and privatization, reflecting a broad shift from Keynesian economics to neoliberalism. The effect of this shift is to center the desires of the richer and more powerful, because they are both the desired consumers and the decision makers within the corporations increasingly making the choices about urban form and change. In this context, women's ascension within the professions potentially gave women more of a voice, but the women who had power, both as decision makers and as consumers, were white and affluent. Like earlier feminists, they often saw their needs as the model of women's needs more broadly.

In the late-20th century, urban design and planning practice, influenced by the ideas of Jane Jacobs, William Whyte, and Jan Gehl, shifted toward mixed uses and mixed populations rather than the separation and sorting of modernist planning. New Urbanism, which sought to reproduce the space and mixture of older cities to promote sociability, used urban design to similarly promote mixture. This shift in ideals bode well for breaking down the barriers between domestic and public, but single-use suburban developments still remained the norm. In addition, the ideal city as understood by Jacobs, Whyte, Gehl, and the New Urbanists was still built primarily on the experiences of white, often male, residents. For example, Whyte's studies of small urban spaces examined the use of spaces without addressing gender differentials in space use, and as the majority of users were men, his recommendations reinforced the masculinity of space, prioritizing people watching over defensible space (Mozingo 1989). The raced and gendered underpinnings of the historic urban structures mimicked by New Urbanists were never questioned; the past was seen as positive in historicist practice. Urban dwellers were relatively undifferentiated by scholars and designers who saw any increase in interaction and use of public space as positive and ignored the problems of street harassment of women, queers, and people of color and did not critically address who did not use urban spaces and why.

2000–present: addressing the structures of inequality

The early-21st century is marked by a greater understanding of both the structural nature of inequality and the ways that gender intersects with other structures of difference. While many planners and activists in the late-20th century argued that increasing the numbers and roles of women in planning and other design and policy professions could transform cities, experience has shown that the issues facing planning are more complex than just representation. In part, the persistence of the glass ceiling, supported by the masculinist structure of the workplace, has kept many women out of top decision-making positions and pushes many women out of allied design professions.[2] The revelations of the #MeToo movement made visible the deep misogyny and masculine privilege operating in many workplaces, while the frustrations of parents, disproportionately women, who have been sidelined by workplaces' lack of flexibility to accommodate childcare, provided further evidence that it is not enough for women to enter the workplace unless workplace norms change dramatically. Underlying the limitations of representation are the structures of power, both in the workplace and in the city, which have increasingly become the focus of both feminist activism and planning work that strives for gender equity. We are beginning to analyze the tools of planning in order to understand how they embody and enact gendered and other structures of power, for example in their definitions of work and of expertise. Even simple norms such as the timing of public planning committees and community events, which are often organized to avoid 9-to-5 working hours but are not sensitive to the scheduling needs of parents, are beginning to be called out for their prejudicial effects.

At the same time, feminists' understandings of women's needs have transformed. Critiques of the way that feminism is too often based on white middle-class experience have been part of the feminist movement from its beginnings, but these critiques have begun to be taken more seriously recently in the context of nearly constant awakenings to the reality of race-based violence, the reinvigoration of overtly racist movements, and thoughtful unveilings of the nature of white privilege. Academic theories from the early 1990s on the social construction of both gender and race, as well as Black feminist writings from the same era on the intersections between race and gender, have become increasingly mainstream tools to understand the limits of inclusion and the persistence of both misogyny and racism. Queer and trans theory have further complicated underlying assumptions about the stability of sexed bodies and critiqued the heteronormativity of understandings of gender. Scholarship has increasingly paid close attention to the differences between women as well as the commonalities, a task taken up in books such as *Gendering the City* (2000), which addresses class, race, and indigeneity in the North American context, and *Cities and Gender* (2009), which brings together discussions of Europe, North America, and the Global South. As the example of Gender Mainstreaming in Vienna shows (see Case study), it is necessary to respond to women's varied needs and to attend particularly closely to the needs of non-white, poorer women to create a city that truly welcomes women.

However, planning that works to address and remediate gendered and other urban inequalities has been limited because of the continued power of neo-liberal planning. New projects are expected to pay for themselves, and the persistence of public-private partnerships, conservancies, business improvement districts, and privately owned public spaces decrease public oversight and magnify the influence of corporations. Not only are these corporations usually controlled by elite white men, but the demand for profitability also centers the desires of elite white men as users, as they have the most money to spend. In this context, gender equity needs to be presented as an asset, a plus that improves profitability, rather than being accepted as a human right. The needs of poor women, especially women of color, are largely unmet when they are seen as a drain on profitability.

Conclusion

Women have been part of cities since their beginnings, but planners have tended to shape cities as spaces for men. Extrapolating from their own experience, they have created zoning and transportation systems that may work well for the isolated individual worker, but make life more difficult for women, who are often struggling with the second shift of household and childrearing labor. Planners have worked to make the city orderly and have treated domestic space as a respite from the public world of work. In so doing they have made women's trips between places beyond the poles of home and work onerous and have made it more difficult to integrate household labor and waged work. The idea that women are endangered by the social mixture and contact of the city, and that safety is their primary need, has led to urban decisions that restrict women. As women have demanded their right to make decisions as planners and policy makers and have asserted their right to the public spaces of the city, whether as shoppers, workers, political activists, or simply as humans, new planning ideas that begin with the needs of those at the margins, including women, are gaining ground.

As we move forward, we need to question the traditional assumptions underlying how we plan transportation, land use, and other aspects of cities, always checking to see what gendered assumptions are built into our conventional practices and actively making sure that all women's experiences are taken into account. In addition, we need to look beyond the traditional tools of planning to contest how patriarchal ideology underlies other systems, structuring the norms of the workplace, the funding of public services, the ways we define the economy, and banking, for example. We could start by interrogating the ways that a household is defined by the census, including the designation of a single "householder" (previously "head of household") in relationship to whom all other members of the household are understood. We can rethink the way that only paid work is considered valuable by most economic models, discounting the value of care work done primarily by women and undermining broad support for childcare as a right. We can challenge the many barriers that push women, particularly women of color, out of the built environment fields because of white-male-centered models of excellence and fit and workplace norms that assume a caregiver at home.

Alternative planning history and theory timeline

Period	Planning history and theory canon	Alternative timeline: women
Late 1800s–1914	Birth of planning	Women claim the city New feminized shopping districts, settlement houses, YWCAs, women's civic clubs, young working women
1914–1945	Formalization of planning	Modern working women Birth of sexist/racist zoning, unions, women professionals
1945–1965	Growth of planning	Redomestication Suburban isolation and sorting by class and race, racist/sexist redevelopment
1965–1980	Midlife crisis of planning	Spaces of women's liberation Women's centers, Take Back the Night, legal challenges to all-male space
1980–2000	Maturation of planning	Feminizing the city More diverse planners, planning for safety, gentrification, neoliberalism
2010–present	New planning crisis	Addressing the structures of inequality Intersectionality, queer critiques, #MeToo, Gender Mainstreaming, community-led planning

Case study: Gender Mainstreaming of urban planning

Gender Mainstreaming, the practice of ensuring that women's perspectives and gender equality is at the center of all decisions, as pioneered in planning by the city of Vienna, Austria, has recently become more prominent in planning discussions within the Anglosphere. In 2019, the American Planning Association put out a white paper that sets out a framework for integrating Gender Mainstreaming into planning in the United States, using specific examples based in transportation planning (Ryan 2019). While this approach has not yet been integrated into US planning, there has been a groundswell of support for instituting Gender Mainstreaming in Toronto, Canada. In addition, The City of Edmonton in Canada has instituted a Gender-Based Analysis Plus and

Equity toolkit that has been used to guide Edmonton's zoning code revision – with results including strong support for accessory buildings or "granny flats" (Edmonton; Klingbeil, 2021). However, continental Western Europe has been more advanced than the Anglosphere and other parts of the world in this respect. In Vienna, Gender Mainstreaming has been at the core of planning decisions since 1992. In order to plan the city in a way that serves all inhabitants in an equitable way, planners in Vienna counter the masculinist bias of traditional planning practice by differentiating users by gender, life phase, and other aspects of difference and centering the needs of those who tend to be underrepresented in planning processes, particularly women, the elderly, and young children. They apply the principles of supporting the compatibility of family duties and paid work, equitable distribution of spatial resources, safe and attractive living environments, and equitable participation in development and decision-making. In order to meet these goals, all plans and designs are evaluated for the effects they have on all user groups. The *Manual for Gender Mainstreaming in Urban Planning and Urban Development* (Urban Development Vienna 2013) provides examples of using Gender Mainstreaming in every aspect of the planning process, including master planning, urban design, land use, development planning, public space planning, public buildings, and housing. The Gender Mainstreaming planning process begins with surveying a wide variety of users about their needs. For example, how do males and females of different ages use public transportation? What variety of activities do they do in a typical day, and in what places? What kinds of spaces do they use in parks? Research on mobility, for example, found that women in Vienna take more trips to shop, take care of family chores, and accompany children, and that more of women's trips involved walking and public transportation. These findings led to a set of planning objectives related to mobility, focused on supporting a switch away from the use of privately owned cars. This included design that prioritized pedestrians and creating pleasant, spacious, safe walking paths; design that promoted cycling; plentiful public toilets; user-friendly public transport, including good service outside peak hours; and mixed-use zoning that helped to minimize the length of trips. Any proposed urban design can be evaluated on this basis, in part by doing an everyday route check that maps out the routes taken on a typical day by a wide variety of different users. These design norms clearly fit with goals of sustainable and lively cities, showing how centering the needs of women leads to design that serves everyone better. While several broad surveys underlie the creation of broad principles for Gender-Mainstreamed design, each new project in Vienna includes further focused research and input from those usually left out of conventional planning. As the processes and principles underlying Gender Mainstreaming have matured and gained influence, Viennese planners have recognized that focusing on gender is necessary but not sufficient for creating fairly shared cities. Hence, they have broadened their focus on the intersecting identities and needs of those marginalized by traditional planning.

Notes

1 The Athens Charter was a very influential document of urban planning principles written by Le Corbusier and based on the 1933 Congrès International de l'Architecture Moderne (CIAM), which was focused on urbanism.
2 While women make up approximately 40% of planners in the United States and United Kingdom, the gap between male and female wages is 20%–29%, in part reflecting fewer women in high management positions. In architecture, women make up 25% of the profession in the United States, 29% in the United Kingdom and Australia, and 30% in Canada. However, women are significantly less likely to be registered architects (17% in the United States, 24% in Australia) and it has been estimated that 32% of women architects in the United States leave the profession. Statistics from U.S. Census, Royal Town Planning Institute (United Kingdom), American Institute of Architects, Architects Registration Board (United Kingdom), The Association of Collegiate Schools of Architecture (United States), Parlour (Australia), and Equity By Design (United States).

References

Deutsch, Sarah. 2000. *Women and the City: Gender, Power, and Space in Boston, 1870–1940.* Oxford: Oxford University Press.

DuBois, Ellen Carol and Lynn Dumenil. 2009. *Through Women's Eyes: An American History with Documents.* 2nd ed. Boston: Bedford/St. Martin's.

Edmonton, Canada. 2021. "Equity and the Zoning Bylaw." https://www.edmonton.ca/city_government/urban_planning_and_design/equity-and-the-zoning-bylaw accessed 20 March 2021.

Enke, Anne. 2007. *Finding the Movement: Sexuality, Contested Spaces, and Feminist Activism.* Durham: Duke University Press.

Enstam, Elizabeth York. 1998. *Women and the Creation of Urban Life: Dallas, Texas, 1843–1920.* College Station: Texas A&M University Press.

Flanagan, Maureen A. 2018. *Constructing the Patriarchal City: Gender and the Built Environments of London, Dublin, Toronto, and Chicago, 1870s into the 1940s.* Philadelphia: Temple University Press.

Friedan, Betty. 1963. *The Feminine Mystique.* New York: W.W. Norton.

Greed, Clara H. 1994. *Women and Planning: Creating Gendered Realities.* London: Routledge.

Gutman, Marta. 2014. *A City for Children: Women, Architecture, and the Charitable Landscapes of Oakland, 1850–1950.* Chicago: University of Chicago Press.

Jacobs, Jane. 1961. *The Death and Life of Great American Cities.* New York: Random House.

Jarvis, Helen, with Paula Kantor and Jonathan Cloke. 2009. *Cities and Gender.* London: Routledge.

Klingbeil, Cailynn. 2021. "Edmonton is Making Its Alleyways a Great Place to Live" *Next City* 1 February 2021. https://nextcity.org/daily/entry/edmonton-is-making-its-alleyways-a-great-place-to-live.

Matrix. 1984. *Making Space: Women and the Man-Made Environment.* London: Pluto Press.

Meyerowitz, Joanne. 1991. *Women Adrift: Independent Wage Earners in Chicago, 1890–1930.* Chicago: University of Chicago Press.

Miranne, Kristine B. and Alma H. Young, eds. 2000. *Gendering the City: Women, Boundaries, and Vision of Urban Life.* Lanham, MD: Rowman and Littlefield Publishers, Inc.

Mozingo, Louise. 1989. "Women and Downtown Urban Spaces." *Places* 6:1, pp.38–47.

Peiss, Cathy. 1986. *Cheap Amusements: Working Women and Leisure in Turn-of-the-Century New York*. Philadelphia: Temple University Press.

Roberts, Marion. 1991. *Living in a Man-Made World: Gender Assumptions in Modern Housing Design*. London: Routledge.

Ryan, Sherry. 2019. Integrating Gender Mainstreaming into U.S. Planning Practice. American Planning Association Planning Advisory Service.

Sewell, Jessica Ellen. 2011. *Women and the Everyday City: Public Space in San Francisco, 1890–1915*. Minneapolis: University of Minnesota Press.

Spain, Daphne. 2001. *How Women Saved the City*. Minneapolis: University of Minnesota Press.

Spain, Daphne. 2016. *Constructive Feminism: Women's Spaces and Women's Rights in the American City*. Ithaca: Cornell University Press.

Stimpson, Catherine R., Elsa Dixler, Martha J. Nelson, and Kathryn B. Yatrakis, eds. 1980. *Women and the American City*. Chicago: University of Chicago Press.

Urban Development Vienna. 2013. *Manual for Gender Mainstreaming in Urban Planning and Urban Development*. https://www.wien.gv.at/stadtentwicklung/studien/pdf/b008358.pdf.

Wilson, Elizabeth. 1991. *The Sphinx in the City: Urban Life, the Control of Disorder, and Women*. Berkeley: University of California Press.

Wood, Sharon E. 2005. *The Freedom of the Streets: Work, Citizenship, and Sexuality in a Gilded Age City*. Chapel Hill: University of North Carolina Press.

Suggestions for further study

- Fainstein, Susan S., and Lisa J. Servon. 2005. *Gender and Planning: A Reader*. New Brunswick: Rutgers University Press.
- Kern, Leslie. 2020. *Feminist City: Claiming Space in a Man-made World*. London: Verso.
- Women and Environments International Magazine. http://www.yorku.ca/weimag/.

2

THE POOR

Contested spaces of deprivation and homelessness

Ella Howard

Introduction

Urban planning as a profession emerged in the United States as part of the Progressive-era reform movement. This origin imbued the field with some orientation toward ameliorating social ills, but the relationship between urban planning and poverty has not always been linear or apparent. The goal of urban planning, according to Frederick Law Olmsted, Jr., one of its founders, was to "exert a well-considered control on behalf of the people of a city over the development of their physical environment as a whole" (Peterson 2003, 2). Previously, aspects of American cities had been consciously planned, such as Savannah's famous system of squares, but no modern effort had been made to professionalize and rationalize urban design in a comprehensive manner.

Urban planning and development efforts in the United States have often prioritized the interests of capital. It is a truism, for instance, that maximum profits are rarely, if ever, achieved by urban real estate developers who choose to design housing for the middle classes, let alone the working classes and the poor. Some planners have envisioned ways to improve the lives of the poor, but others have carried out plans and developments designed to remove the poor and homeless from urban areas desired by the upper classes. The homeless emerge here as the poorest of the urban poor, appearing in planning and development discussions most often as a "problem," rather than as subjects worthy of a professionally designed environment.

Situating a historical analysis of the ways in which urban planning has addressed the issues of poverty and homelessness is complicated by the fluidity of the history of the field. In the interest of breadth, this chapter adopts an expansive view of the field, focusing on the areas of the history of planning and urban development that have had a significant impact on the lives of the poor.

DOI: 10.4324/9781003157588-3

Late 1800s–1900: urban poverty and visionary reformers

As urbanist Peter Hall has emphasized, the earliest urban planners were radical social reformers who sought to ameliorate the lives of city residents. He argued, "Twentieth-century city planning, as an intellectual and professional movement, essentially represents a reaction to the evils of the nineteenth-century city" (Hall 2014, 7). Early planning efforts developed alongside other social reform movements, sharing a commitment to Progressive ideals.

As urban work opportunities emerged in manufacturing, retail, and other industries, prospective laborers came to the cities from rural areas of the United States as well as other countries, such as Ireland, Italy, and eastern Europe. Upon moving into urban centers like New York City, Boston, and Chicago, they found themselves crowded into tenement buildings. As landlords sought to maximize their profits, the living spaces allotted to each family shrank. Compounding this issue, many families also carried out piecework for area businesses inside their apartments, sewing or rolling cigars amidst their living space.

Seeing an opportunity to help the impoverished immigrant communities social activists drawing inspiration from the English settlement house movement, launched an American version. The Neighborhood Guild was established in 1886 on the Lower East Side of Manhattan. Jane Addams and Ellen Gates Starr soon followed, launching Chicago's Hull House. The movement spread quickly across the country, although each house developed its own targeted mission, with some more focused on vocational training and others on domestic support. The settlement house movement planted the seeds of the field of social work as well as broader public programs (Lissak 1989; Schwartz 2000).

Other reformers tackled the structural issues surrounding housing in particular, using legislation to change regulations on density and features such as air shafts and windows. Jacob Riis, an immigrant photojournalist, used his skills and platform to raise awareness of the plight of those enduring substandard living conditions on the Lower East Side. His 1890 book, *How the Other Half Lives*, incorporated dramatic photographs and lurid descriptions of the homes of area residents. His work inspired further tenement reform legislation and Progressive activism on behalf of the urban poor.

The rise of American cities also caused an increase in the number of homeless individuals, as transient workers followed railroads and employment opportunities. Police stations had traditionally offered temporary lodging to transients, who also frequented religious missions sponsoring meals. As homelessness increased, so did moralistic judgment of the poor. The Charity Organization Society operated services such as their Woodyard in New York City, which sold books of tickets to supporters. When approached by a homeless individual, one could give them a ticket, which they could redeem for the opportunity to chop wood in exchange for a meal and a night's lodging. This program was marketed as a way to avoid fueling the indiscriminate support of the "undeserving poor," a vestige of the English Poor Law. Only the truly needy who were willing to work for their support would therefore receive aid.

Districts welcoming the homeless, known as skid row areas, began to form in many cities in the late 19th century, providing middle- and upper-class Americans a way to avoid interacting with, or even seeing, the poorest of the poor. Such districts were home to low-cost hotels, simple restaurants, and bars with permissive policies. The accommodations they provided would remain limited, in part reflecting the desire of many Americans never to reward the poor with comfort or an aesthetically pleasing life. The spaces of homelessness were often (and continue to be) designed as consciously Spartan in order to encourage individuals to desire something different than their current situation (Howard 2013).

The overcrowded and unsanitary conditions of late-19th century cities were perceived by many as a national disgrace. Changing aspects of such a complex and interwoven situation proved difficult, however. The 1893 Chicago World's Fair offered a striking model of a different approach to urban design. The Fair's major buildings, featuring neoclassical architectural styles, were arranged elaborately. Formal and grand promenades linked the buildings, offering visitors the opportunity to experience the wide boulevards one would find in Paris. Urban planning drew inspiration from the Fair, developing the City Beautiful movement around the key themes of rational planning with attention to aesthetics and formal beauty. Inspired by the City Beautiful movement, planners and architects completed many publicly-oriented buildings such as civic centers, public libraries, and post offices in neoclassical architectural styles. Parks and parkways incorporating natural elements and vistas were also central to the movement (Hess 2006).

The City Beautiful movement has been criticized for its lack of attention to social issues including poverty (Hall 2014). Even housing as a broader category was not a focal point for the movement. Planners during this era viewed housing as the purview of the private sector. The poor were seen as beneficiaries of the public betterment provided by the movement. The parks, parkways, boulevards, and public buildings, being accessible to all residents, were seen as a type of uplifting public good. The grand buildings themselves were seen as monuments to the important role of government and the public sector. These early efforts at urban planning set the stage for a more elaborate development of the field.

1900–1945: economic crisis and bureaucratic reform

The City Practical movement that followed the City Beautiful movement is sometimes framed as the birth of the field of urban planning in the United States. Earlier histories overstated the contrast between the City Practical and City Beautiful movements, echoing the sentiments espoused in the 1920s. As the new movement dawned, its proponents described the City Beautiful movement as overly preoccupied with aesthetics and ignoring social problems. More recent scholarship has added nuance to this discussion, exploring the fusion of aesthetic beauty and functionality in the work resulting from both movements (Pipkin 2008).

The first full-time city planner in the United States, Harland Bartholomew, saw beauty in unity and promoted six primary elements of a city plan: streets, transit, transportation, public recreation, zoning, and civic art (Benton "Just the Way" 2018a). His office created two city plans for St. Louis and one for New Orleans, among other commissions. Bartholomew pioneered the meticulous collection and analysis of data as the basis of urban planning. He was also committed to segregating urban spaces by socioeconomic class and race. His plans zoned Black residents into narrow districts, furthering racial segregation in an effort to maintain high property values for white residents. They sometimes identified Black neighborhoods as in need of rejuvenation, and then proposed the displacement of area residents without care for their fates. Scholar Mark Benton has framed such actions as examples of "administrative evil," a term used to describe actions seen as neutral by their perpetrators that have detrimental effects (Benton "Saving" 2018b).

Early in the 20th century, zoning was seen by some social reformers as a potential tool for improving urban conditions and fostering a healthy and safe living environment for all city residents. In application, however, zoning was quickly used to separate people by race and class. In 1926, the Supreme Court supported zoning in the *Euclid v. Ambler* decision, setting the United States on the path of functional segregation. Zoning has been controversial in the United States because of its potential to blunt the profits that can be achieved through unfettered real estate development. Designating areas of a city to be used for specific purposes, such as manufacturing or commerce, prevents those businesses from operating in residential neighborhoods. Height ordinances, a common feature in zoning ordinances, also limit the ability of developers to maximize profits by adding additional residential units. Ordinances restricting the number of residential units allowed per property have a similar effect, while also ensuring that certain neighborhoods remain inhospitable to those of lesser economic means (Silver 1997; Whittemore 2017, 17–19).

The destruction caused by the Great Depression opened new avenues of opportunity for urban planning and development. The crash of the stock market in 1929 was followed by years of high unemployment and record numbers of foreclosures on businesses and homes. By the time President Franklin D. Roosevelt took office in 1933, the nation's economy was teetering on the brink of collapse. The administration moved swiftly to shore up the banking industry, closing the banks for inspection of assets before allowing them to reopen and backing them with the Federal Deposit Insurance Corporation. The administration also set out to stabilize the mortgage lending system by establishing the Home Owners' Loan Corporation (HOLC). The HOLC partnered with local officials to carry out an extensive, carefully documented series of surveys designed to capture a wide range of details about urban neighborhoods, the houses within them, and their residents. The surveys used a standardized set of forms, which asked for details about the materials used to construct each area's homes, the condition of the buildings, and the area's geographic advantages. The forms also asked for detailed information on the area's residential demographics, including data about socioeconomic class, race, and national origins.

The surveys were used to determine mortgage eligibility for homes in the area. Those areas that were deemed advantageous for mortgage lending were coded green on the maps that were produced, while those areas categorized as dangerous for lending were coded red. Areas that were judged to be of levels between red and green were coded yellow and blue. The data about the structural merit and condition of an area's buildings was used only minimally when determining the area classifications. Instead, the demographic information about the area's residents was heavily weighted, with most areas home to Black and Jewish people coded red.

This process of "redlining" positioned already underserved and underprivileged districts for decades of underinvestment, guaranteeing that subsequent generations of residents would be unable to qualify for needed funding to purchase and upgrade homes and businesses. Residents of dilapidated, primarily white neighborhoods were less able to access loans than their counterparts in more prosperous areas. For Black homeowners, these trends proved especially destructive, as even those who were able to purchase homes did not see a return on their investment comparable to that enjoyed by homeowners in predominantly white neighborhoods (Woods 2012; Yamahtta-Taylor 2019).

The onslaught of the Depression exacerbated the problems faced by the nation's homeless. As businesses struggled and implemented cost-cutting layoffs and, ultimately, closures, unemployed individuals increasingly defaulted on their home loans and became unable to pay rent. The newly homeless population flooded into the nation's cities in search of relief. There, they encountered a patchwork system of private and public relief that had been sufficient to meet some basic needs of the mostly single, homeless men of the 1920s. Flophouse single-occupancy hotels, greasy diners, and religious missions were not designed to accommodate the women, children, and families who had been recently displaced. The Roosevelt administration established the Federal Emergency Relief Administration, which provided funding to the Federal Transient Program. The FTP operated similarly to the migrant labor camps of the era, effectively shifting transient homeless people without local affiliations to congregate shelters designed specifically for their use. Although a short-lived program in operation only from 1933 to 1935, the FTP represented a major federal intervention in the care for homeless people.

1945–1965: federal housing programs and urban development

The federal public housing and urban renewal programs of the postwar era also emerged from New Deal reforms. The programs stemmed from an administrative goal of shifting the uses of prime downtown real estate that was home to poor and working-class neighborhoods. By relocating those communities to publicly-funded buildings located in less desirable areas, downtowns could be revitalized.

Boston's West End was one of the first neighborhoods to fall to the wrecking ball of urban renewal, and one of the only projects to displace a primarily white population. The process of urban renewal began when cities requested urban renewal funds and carried out detailed studies of the targeted neighborhoods. As was the case with the earlier HOLC surveys, the process of assessing and documenting the nature and condition of the buildings could be subjective. In the survey of the West End, a working-class neighborhood, data for the nearby, deteriorated skid row area of Scollay Square contributed to the rating of the area as dilapidated and worthy of renewal. Area residents were provided only incomplete information about the pending plans for radical destruction. The area was flattened, and the ensuing developments include the upscale Charles River Park and the Government Center complex. In New York City, Lincoln Center and Stuyvesant Town similarly displaced earlier communities. This first wave of drastic urban renewal prompted pushback leading to later, more nuanced iterations of the program which focused much more on rehabilitation of structures and did not require demolition of existing buildings. The legacy of urban renewal would be one of displacement, increased racial and economic segregation, and decreased trust in government planning initiatives in working-class and poor communities.

Urban renewal had been designed as the compliment to an equally ambitious federal public housing program. Temporary housing was built under the Housing Act of 1937 and expanded programs emerged from the Housing Act of 1949. Coupled with plans for slum clearance and urban renewal, public housing was planned as a way to relocate the urban poor living in substandard housing. Early public housing was highly desirable, featuring modern appliances and well-maintained facilities. Competition to gain admission to some developments was fierce. Selected applicants were required to have a substantial income, and the earnings of public housing residents often exceeded that of other area residents. The funding model for public housing allocated federal funds for construction but required maintenance costs to be funded through rents collected from tenants. Welfare advocates and progressive politicians pushed for policy changes in the 1960s that routed more single mothers and financially destitute families into public housing. As rents fell, they proved inadequate to fund upkeep of the properties. Many families who were placed in public housing also required social services that were not readily available. As a result, public housing slid into a state of physical disrepair and social disruption (Vale 2013).

The public housing projects of this era were no longer the low-rise, low-density neighborhoods built in the early postwar era. Increasingly, public housing took the form of high-rise Modernist buildings. However, the Pruitt-Igoe complex in St. Louis and the Cabrini-Green Homes in Chicago, both of which were ultimately destroyed to great fanfare, came to represent the American system of public housing. They seemed designed to pack as many residents as possible into a narrow footprint, conserving space in more desirable areas for city residents of means. Consolidating the most financially vulnerable families into massive complexes without adequate social supports had immeasurable, negative effects that would last for decades. Not only were individual families subjected to social trauma, but also, the reputation and very concept of public housing became associated with severe stigma and lost the popularity it had enjoyed in the postwar years.

The postwar era saw the American economy enter a period of overall prosperity. The demand for single-family homes prompted waves of new residential construction, including a broad expansion of suburban development. In an era of widespread prosperity, the poorest individuals were viewed as increasingly aberrant. On skid row, the urban homeless, most of whom were single white men through the 1960s, were approached by service providers and politicians through the lens of addiction. Homeless men were seen by many as end-stage alcoholics, no longer able to function within mainstream society. Their physical presence in the city was seen as a problem to be managed, if not solved. The services provided to the homeless focused narrowly on substance use, whether offered through city municipal shelters or religious organizations such as the Salvation Army. Some facilities were built outside of city centers, such as New York City's Hart Island compound, which welcomed homeless alcoholics into a program of detox and rehabilitation on an island starkly separated from city life that was also the burial site of the city's unclaimed dead. The homeless were thus isolated from the city during a period when many other city residents were experiencing financial prosperity.

1965–1980: suburbanization and gentrification

American culture in the 1950s and 1960s was largely defined by suburban, car-oriented life. The suburban growth of the postwar period eclipsed dramatically the earlier iterations, such as the streetcar suburbs of the 1920s. This trend would continue through much of the 20th century, fundamentally altering American social and political life. Several factors coalesced to make possible a large-scale shift toward suburban living. The U.S. military rewarded WWII veterans and their families with attractive mortgage loan programs that allowed buyers to purchase homes with low or no down payment. Developers used low-cost construction techniques to build new suburban developments quickly and inexpensively. Most famously, Levitt & Sons created several developments (Levittowns) in locations like Long Island and Pennsylvania. These new suburban developments were marketed to young families seeking space to raise children and the amenities of a residential neighborhood, including convenient access to schools, supermarkets, and shopping centers (Jackson 1985).

Middle-class Americans flocked to such suburban communities, leaving behind the old apartment buildings and dense downtown districts of their parents' generation. As they did so, they participated in "white flight," a trend that accelerated in the 1960s, as middle-class white Americans gained access to the new suburban neighborhoods. Such access was limited by not only financial means. Racial covenants, which prohibited sale of property to Black families, were incorporated into the legal documentation of some suburban communities. In some others, the climate was so hostile and marked by racist behavior that Black people were made to feel unwelcome. The growth of suburbanization thus led to deeper economic and racial segregation of American cities, as white families with middle-class or upper-class incomes moved to the suburbs.

Amidst the widespread prosperity of the postwar period, American rediscovered urban poverty in the 1960s. Early in the decade, political scientist Michael Harrington's *The Other America: Poverty in the United States* shined a spotlight on the scope of the era's poverty. Drawing on government metrics, Harrington described one-fourth of Americans as living in poverty. The structure of modern cities and suburbs, he argued, effectively hid most of this economic deprivation and suffering from the middle and upper classes, who lived in areas that lacked socioeconomic diversity (Harrington 1997).

Theoretical interpretations of the causes of such poverty were varied. Harrington cited the evolution of the modern American welfare state, which hinged upon Social Security, a limited federal program. Anthropologist Oscar Lewis tried to explain the status of the poor by cataloguing their responsive behaviors. His "culture of poverty" thesis became a convenient weapon for those looking to blame the poor for their plight. When President Lyndon Johnson launched the War on Poverty in 1964, he envisioned a new path forward for American government. A wide range of Great Society programs were established which improved the living conditions of many of the country's most vulnerable residents. The Head Start early childhood education program, Food Stamp Act, and Social Security Acts of 1965, as well as the establishment of the Department of Housing and Urban Development, ensured federal attention and spending would be directed toward poverty issues.

Perhaps the most maligned of these initiatives, the Model Cities program created over 150 five-year experimental programs. The fundamental goal of the Model Cities program was to engage members of underserved communities in municipal government while developing new antipoverty programs. The program reflected a second attempt at the War on Poverty, the first of which had included the Community Action Programs created by the 1964 Economic Opportunity Act. That Act had specifically focused on "maximum feasible participation of members of impoverished communities." While Johnson, arguably the nation's most progressive president, had hoped for a wide-ranging, comprehensive reform of municipal politics, the administration of his successor, Republican President Richard Nixon, took the program in a different direction. Rather than focusing on community participation, Model Cities came to represent programs focused on the construction of buildings and housing reforms. Some neighborhood rehabilitation projects were carried out, but the program did not make the large-scale improvements originally envisioned. Scholars have shown that public and academic perceptions of the Model Cities program have been skewed over time. While the program has often been characterized as a failure, its success in engaging members of marginalized communities in the work of government has been largely overlooked (Weber and Wallace 2012).

Since the 1960s, scholars have analyzed the phenomenon of gentrification. The process has been identified as early as the 1940s, in varying formats. In the postwar period, as American cities wrestled with an out-migration of more affluent individuals and families, some of the areas they left behind fell into disrepair. The lower occupancy rates drove down the costs of rental units and property alike.

In the classic model of gentrification, a group of new residents enter a neighborhood. Often artists or intellectuals, they are not wealthy, but enjoy city life and welcome demographic diversity. Their presence changes the neighborhood in subtle and dramatic ways. New businesses open to cater to their interests and preferences. Older businesses that were more suited to the area's working class residents start to close. The next wave of neighborhood residents tends to have higher status jobs and more discretionary income. They are drawn to the neighborhood, which now features interesting shops and restaurants, and is home to residents from a diverse range of socioeconomic and racial backgrounds. More upscale businesses are established to meet the tastes of the newly arriving residents, who also modify their properties, carving out renovations that increase property values in the area. These forces increase the cost of living to such an extent that the neighborhood's original residents are displaced (Zuk et al. 2018).

The process described above rarely unfolds neatly as written. Variations on the gentrification model have occurred in cities across the country, however, changing the fundamental character of a wide variety of neighborhoods. Urban planners, activists, and others seeking to mitigate the effects of gentrification face a challenging situation. Real estate developers, many homeowners, business owners, politicians, and other interested parties embrace wholeheartedly the core elements of gentrification that lead to neighborhood improvement. Higher property values, an influx of small businesses, and greater investment in a neighborhood are seen by many as desirable outcomes. The more controversial aspect of gentrification is residential and business displacement. Whether quantitative or qualitative in nature, much scholarly and professional research has attempted to understand the process of gentrification in order to blunt its negative effects.

1980–2000: housing and shelter in the modern city

New Urbanism, a design movement that has been influential since the 1990s, focused on developing livable cities and combating urban sprawl. New urbanists promoted a return to residential density and supported walkable neighborhoods. Responding to climate change, they also urged environmentally responsible practices. The movement also called for attention to local history and culture, to ensure that each neighborhood would reflect its unique environment, rather than replicating a generic model in cookie cutter fashion (Hanlon 2010). These principles were incorporated in the Hope VI program, which launched in the early 1990s. Administered by the

United States Department of Housing and Urban Development, the program carried out a major reform of public housing. Hope VI set out to replace earlier public housing developments with new mixed-income ones. The program did not require a direct substitution for demolished public housing units.

The program has been criticized for relocating some low-income families from housing developments and leaving them displaced from their communities. A mixed-finance approach also undergirded the program, which was funded by various public and private entities, relieving the federal government from responsibility for funding an entire development. The program's emphasis on creating mixed-income and mixed-finance developments led to attention instead to the needs of families and individuals with means. Despite such criticisms, studies have found positive outcomes for some displaced families who were issued housing vouchers to rent in areas with less concentrated poverty and more social and labor opportunities. However, the program has not fostered racial integration in any significant way. Because of its limited approach to social support services, the program has also not been able to assist some of the nation's most vulnerable families (Popkin et al. 2009).

Beginning in the 1970s, however, cities across the country witnessed new trends in poverty and homelessness. As overall poverty rates started to rise, more homeless women began to appear. This "new homelessness" resulted from a complex matrix of related factors, but average observers struggled to understand the trend. Media outlets shared panicked and sordid accounts of "bag ladies," connoting an alien culture. Homeless women began to appear in New York City in large numbers in the 1970s, reflecting cuts to social services, rising rates of deinstitutionalization of the mentally ill, and shifting housing costs. The numbers of homeless women in Boston would rise to 2,000 by the mid-1980s. In 1974, the nation's first homeless shelter for women opened in Boston's South End when community activist Kip Tiernan founded Rosie's Place.

During the 1980s, American homelessness changed in dramatic ways. Women and children, always vulnerable to assault, began to appear more frequently on the streets and in the shelter system, as did more young, Latino, and Black men. Fewer skid row areas were available, as many had fallen either to urban renewal or gentrification. As more homeless individuals and families spread into visible areas, their presence was increasingly criminalized. Vagrancy laws, which had emerged after the Civil War as a method of restricting the behavior of newly freed Black Americans, were enforced in cities and suburbs alike. "Right to Shelter" laws passed in New York and other cities (states) guaranteed homeless individuals accommodations on demand, but need often exceeded available resources.

2010–present: the forces of capital and care

In the past decade, a series of environmental crises have been having disproportionate effects on low-income and homeless individuals. As demonstrated by the 2005 devastation of New Orleans by Hurricane Katrina, a city's poorest residents are often clustered in the least desirable and most dangerous areas. In the years since the storm, few impoverished former city residents have been able to return to the city, finding themselves priced out by households and investors with more resources. Similarly, the lack of clean water in Flint, Michigan exposed nationally in 2015, revealed the disparate effects of an environmental crisis on households of varied means. From April 2014 through October 2015, the city drew its drinking water not from its customary Detroit sources, but from the Flint River. The water consumed by area residents was proven to be contaminated with lead and possibly other dangerous materials. Most residents of Flint are Black and 39% live below the poverty line. The city's median household income in 2015–2019 was below $30,000. As had been the case in New Orleans, low-income Black communities struggled to get politicians to see their needs as vital. Eventually in this case, activists generated outrage and national and global support, securing state and federal resources to remedy the situation.

More broadly, impoverished communities are more dependent on public infrastructure than their wealthy counterparts. Access to bottled water in case of contamination, cars when public transportation is interrupted or unavailable, and the internet during a global pandemic all require financial resources that many Americans do not have (Pauli 2019). Moreover, the global financial crisis that began in 2007 triggered a subprime mortgage crisis in the United States. When housing values declined from record highs, many Americans found themselves unable to make their scheduled mortgage payments. A wave of foreclosures and evictions left many families and neighborhoods in ruin. Private equity firms and landlord-investors seized upon this opportunity, buying up large numbers of properties and leveraging them for maximum profits through rental or resale (Seymour and Akers 2019).

The contemporary landscape of Boston's South End highlights the intersection of the forces discussed in this chapter. In 2007, the *Boston Globe* observed, "Once a gritty slum of townhouses and brownstones, the South End has been transformed over the last quarter of a century into one of Boston's most coveted neighborhoods, with condominiums that routinely sell for more than $1 million" (Badkhen 2007). Yet for all these changes, the Pine Street Inn shelter remains a neighborhood fixture, assisting nearly 2,000 homeless individuals per day. Further down Harrison Avenue, Rosie's Place community center welcomes homeless and poor women, offering an array of services, including a day shelter, meals, a food pantry, legal advocacy, and English as a Second Language classes. The term "skid row" is rarely used these days but the stretch of Massachusetts Avenue in the South End referred to as "Methodone Mile," reflects the visible presence of shelter and opioid treatment facilities, as well as their clients (Zalkind 2017).

New York's skid row, the infamous Bowery, had survived urban renewal only to fall to gentrification. As homelessness spiked in the 1980s, in part due to policies of austerity, homeless individuals began to seek shelter across the city, often in the outer boroughs. New York City Mayor Rudolph Giuliani's administration's strict policing policies further ensured the homeless were less visible in the urban center, with vigorous patrolling in desirable residential and commercial neighborhoods. Boston, on the other hand, cultivated the South End as a skid row district for decades. Aside from periodic residential complaints, this posed relatively few problems until the rise of the opioid epidemic in the mid-2010s. Now, the population of drug-addicted individuals is pressed into the same zone as people rendered homeless by other factors. The district is becoming increasingly difficult for city officials and social service agencies to manage.

Similar challenges have plagued the Housing First strategy, which promotes housing as the logical solution to the problem of homelessness. While previous "Housing Readiness" approaches to homelessness focused on guiding homeless individuals to sobriety before providing them with housing, the Housing First model concentrates directly on placing clients in housing. The model was espoused by the National Alliance to End Homelessness in 2006 and adopted as a national goal in 2010 by the United States Interagency Council on Homelessness. Homeless clients outpace available housing, however, especially given the need for landlords to approve of tenants being placed in their buildings. In the competitive placement environment, scholars have observed that priority is often given to cooperative clients who align with specific categories of disability and vulnerability that will smooth bureaucratic hurdles toward placement. The least stable and most severely mentally ill, as well as those in the deepest throes of addiction thus often remain unhoused (Osborne 2019).

Many homeless people have gathered in cities with warmer weather and more hospitable policies, such as San Francisco, Seattle, and Los Angeles. While many cities were eradicating their skid row areas in the 1970s, Los Angeles officials were overseeing the consolidation of social services and lodgings downtown. In recent years the expanding homeless population has come into conflict with the rapidly gentrifying surrounding area. By 2020, the City of Los Angeles was home to more than 41,000 homeless people. Many remained concentrated on skid row, while others lived in smaller encampments outside the area. In 2021, Mayor Eric Garcetti called for spending almost $1 billion to address the city's homelessness crisis. Judge David Carter ordered the city to go further, calling for an audit of the billion dollar plans as well as the procurement of shelter for all of the city's homeless by October 2021. These bold and unprecedented plans speak to the extent of the crisis gripping the city.

Conclusion

Throughout the 20th century and beyond, the poor and homeless have been viewed as urban problems to be solved. Responding to the sanitation and crowding issues of the late-19th century city, urban planners created spaces of beauty and reflection. Later generations navigated the space between aesthetics and stark function with varying proposals, at times promoting urban density and at other times cultivating suburbs.

Major governmental interventions brought vast change to the lives of America's urban poor. The creation of public housing improved the quality of life for many families, while redlining and urban renewal left many in despair. Similarly, federal assistance to the homeless provided by the Federal Transient Program and policies such as Right to Shelter and Housing First has created housing opportunities for many homeless individuals and families. The communities developed under the HOPE VI program of the 1993–2010 period offer a forward-looking approach that has helped to partially restore confidence in public housing. Overall, the poor and homeless have benefited from plans developed by urban planners, government agencies, and others during times of crisis, as responses to highly visible unmet need.

The optimism of the early urban planners and the activists of the 1960s seemed misplaced and unrealistic for decades, as public-private growth partnerships, displacement, and gentrification dominated urban agendas. The reform spirit has been revived, however, through the Community Land Trusts (CLTs) (see Case study) developed in recent decades. Projects like CLTs developed through open communication and real cooperation with members of impacted communities can render the poor not mere subjects, but true partners in urban planning.

Alternative planning history and theory timeline

Period	Planning history and theory canon	Alternative timeline: the poor
Late 1800s–1900	Birth of planning	*Urban poverty and visionary reformers* Urbanization, tenement reform, settlement houses, establishment of skid rows
1900–1945	Formalization of planning	*Economic crisis and bureaucratic reform* Zoning and segregation by class and race, New Deal policies, redlining
1945–1965	Growth of planning	*Federal housing programs and urban development* Urban renewal, public housing, alcoholism studies
1965–1980	Midlife crisis of planning	*Suburbanization and gentrification* Suburban white flight, rediscovery of poverty, War on Poverty, displacement
1980–2000	Maturation of planning	*Housing and shelter in the modern city* Hope VI, housing vouchers, rediscovery of homelessness, homeless women and children
2010–present	New planning crisis	*The forces of capital and care* Environmental crises, subprime mortgage crisis, Housing First

Case study: Community Land Trusts

The "American dream" is often framed in terms of owning one's own home and/or business. Echoing the language of Jeffersonian America, many traditional cultural milestones revolve around marriage, homeownership, and having children. During the postwar decades, many white American families benefitted from the expanded mortgage lending programs and steadily increasing property values, making home ownership a primary vehicle for financial gain for middle-class families. Redlining and racially restrictive covenants, coupled with individual acts of outright racism, prevented many Black Americans from enjoying those opportunities. Even when programs specifically focused on fostering homeownership for low-income Black families, racist policies and practices kept property values in those neighborhoods from appreciating in a manner commensurate with the gains seen in predominantly white neighborhoods. Predatory lenders and others who encouraged homeownership without appropriate safeguards led to financial ruin for countless families. An alternative vision for low-income home ownership, Community Land Trusts, arose from the Southern Civil Rights movement. New Communities, Inc. was established near Albany, Georgia, with the goal of securing homeownership for African Americans who had been systematically impoverished by white elites. By leasing portions of a 5,000-acre parcel of land to families for home ownership and farming, families would earn profit and gain home equity. Although the gambit ultimately failed, it inspired others to take up similar efforts. A classic example of a successful Community Land Trust is Dudley Neighbors, Inc., located in Boston's Roxbury neighborhood. Roxbury was known as the heart of the city's Black community by the 1970s, but decades of neglect and racism had left the area in distress. Community activists and area residents launched Dudley Neighbors, Inc. in 1988 and secured the support of the Boston Redevelopment Authority, who extended to them the power of eminent domain, the right to purchase land in pursuit of the public good, regardless of the seller's intent. The Mayor's office and City Council have seats on the Dudley Neighbors, Inc. Board, ensuring open communication and the pursuit of common goals. Through a system similar to a mobile-home park, Dudley Neighbors, Inc. retains ownership of the land, while homeowners own their structure. When they sell their home, the resale price is capped to ensure that the next family to purchase the home finds it affordable. Community Land Trusts vary in structure and area and are designed to meet the needs of particular communities. Some Community Land Trusts determine the resale value of a home by measuring neighborhood incomes. Others use a two-appraisal process to determine the resale value, ensuring that sellers recapture their initial investment, plus a portion of the profit they would have earned selling the home in an unrestricted environment. Some Community Land Trusts have been initiated by municipal government agencies while others operate at arm's length. Many Community Land Trusts also provide financial counseling and education about key issues faced by homeowners, ensuring that community members can make fully informed decisions. Community Land Trusts allow many low-income individuals to achieve the goal of homeownership within a structured and supportive environment. When managed with care, they protect homeowners from predatory lenders and an unpredictable real estate market. Scholars have noted increased interest in Community Land Trusts after 2000. Such interest may signal a path forward for community-based urban planning in the years to come.

References

Badkhen, Anna. 2007. "Say Goodbye, or Good Riddance, to Old Pub." *Boston Globe*, September 15, 2007.

Benton, Mark. 2018a. "'Just the Way Things Are Around Here': Racial Segregation, Critical Junctures, and Path Dependence in Saint Louis." *Journal of Urban History* 44, no. 6: 1113–30.

Benton, Mark. 2018b. "'Saving' the City: Harland Bartholomew and Administrative Evil in St. Louis." *Public Integrity* 20, no. 2: 194–206.

Hall, Peter. 2014. *Cities of Tomorrow: An Intellectual History of Urban Planning and Design since 1880*, Fourth Edition. NJ, USA: Wiley & Sons.

Hanlon, James. 2010. "Success by Design: HOPE VI, New Urbanism, and the Neoliberal Transformation of Public Housing in the United States." *Environment and Planning* 42: 80–98.

Harrington, Michael. 1997. *The Other America: Poverty in the United States*. New York: Touchstone.

Hess, Daniel Baldwin. 2006. "Transportation Beautiful: Did the City Beautiful Movement Improve Urban Transportation?" *Journal of Urban History* 32, no. 4 (May): 511–45.

Howard, Ella. 2013. *Homeless: Poverty and Place in Urban America*. Pennsylvania: University of Pennsylvania Press.

Jackson, Kenneth T. 1985. *Crabgrass Frontier: The Suburbanization of the United States*. New York: Oxford University Press.

Lissak, Rivka Shpak. 1989. *Pluralism and Progressives: Hull House and the New Immigrants: 1890–1919*. Chicago: University of Chicago Press.

Osborne, Melissa. 2019. "Who Gets 'Housing First'? Determining Eligibility in an Era of Housing First Homelessness." *Journal of Contemporary Ethnography* 48, no. 3: 402–28.

Pauli, Benjamin J. 2019. *Flint Fights Back: Environmental Justice and Democracy in the Flint Water Crisis*. Cambridge, MA: The MIT Press.

Peterson, Jon A. 2003. *The Birth of City Planning in the United States, 1840–1917*. Baltimore: Johns Hopkins University Press.

Pipkin, John S. 2008. "'Chasing Rainbows' in Albany: City Beautiful, City Practical 1900–1925." *Journal of Planning History* 7, no. 4 (November): 327–53.

Popkin, Susan J., Diane K. Levy, and Larry Buron. 2009. "Has Hope VI Transformed Residents' Lives? New Evidence from the Hope VI Panel Study." *Housing Studies* 24, no. 4: 477–502.

Riis, Jacob. 1890. *How the Other Half Lives: Studies Among the Tenements of New York*. New York: Charles Scribner's Sons.

Rothstein, Richard. 2017. *The Color of Law: A Forgotten History of How Our Government Segregated America*. New York: W.W. Norton.

Schwartz, Joel. 2000. *Fighting Poverty with Virtue: Moral Reform and America's Urban Poor, 1825–2000*. Bloomington: Indiana University Press.

Seymour, Eric, and Joshua Akers. 2019. "Building the Eviction Economy: Speculation, Precarity, and Eviction in Detroit." *Urban Affairs Review* 57, no. 1 (June): 35–69.

Silver, Christopher. 1997. "The Racial Origins of Zoning in American Cities." In *Urban Planning and the African American Community: In the Shadows*, edited by June Manning Thomas and Marsha Ritzdorf, 23–42. Thousand Oaks, CA: SAGE.

Vale, Lawrence J. 2013. "Public Housing in the United States: Neighborhood Renewal and the Poor." In *Policy, Planning, and People: Promoting Justice in Urban Development*, 285–306. Pennsylvania: University of Pennsylvania Press.

Weber, Bret A., and Amanda Wallace. 2012. "Revealing the Empowerment Revolution: A Literature Review of the Model Cities Program." *Journal of Urban History* 38, no. 1 (January): 173–92.

Whittemore, Andrew H. 2017. "The Experience of Racial and Ethnic Minorities with Zoning in the United States." *Journal of Planning Literature* 32, no. 1: 16–27.

Woods, Louis Lee. 2012. "The Federal Home Loan Bank Board, Redlining, and the National Proliferation of Racial Lending Discrimination, 1921–1950." *Journal of Urban History* 38: 1036–59.

Yamahtta-Taylor, Keeanga. 2019. *Race for Profit: How Banks and the Real Estate Industry Undermined Black Homeownership*. University of North Carolina Press.

Zalkind, Susan. 2017. "The Infrastructure of the Opioid Epidemic." CITYLAB, September 14, 2017. https://www.citylab.com/equity/2017/09/methadone-mile/539742/.

Zuk, Miriam, Ariel H. Bierbaum, et al. 2018. "Gentrification, Displacement, and the Role of Public Investment." *Journal of Planning Literature* 33, no. 1: 31–44.

Suggestions for further study

- Desmond, Matthew. *Evicted: Poverty and Profit in the American City*. Largo, MD: Crown Books, 2016.
- Freidrichs, Chad (director). 2011. *The Pruitt-Igoe Myth: An Urban History*. Documentary film.
- The Lower East Side Tenement Museum. www.tenement.org.
- Patterson, James T. 2000. *America's Struggle Against Poverty in the Twentieth Century*. Cambridge, MA: Harvard University Press.

3

LGBTQ+ COMMUNITIES

Sexuality and the city

Tiffany Muller Myrdahl

Introduction

Normative understandings of sexuality are encoded into our everyday landscapes (Hubbard 2001). In the North American settler-colonial context, expectations of heteronormativity, heteropatriarchy, and compulsory monogamy shaped colonial settlement and continue to inform the development of the urban landscape. Heteronormativity refers to heterosexuality – sexual intimacy between opposite-sex people – as the culturally sanctioned expectation, supported by church and state. Heteropatriarchy assumes the primacy of reproductive heterosexual relationships in the service of maintaining a patriarchal system where decision-making flows from, and inheritance benefits, the male head-of-household. Compulsory monogamy is the state-sanctioned norm that a sexual union is between two (heterosexual) people (Tallbear 2019). In short, specific, normative expectations of sexuality are foundational to the ways that the landscape is organized via private property at the scale of the household and beyond.

While perhaps an unlikely starting place for a chapter on sexual- and gender-diverse communities, it is essential to ground the discussion of these communities with pointed attention to the role that normative sexual values play in the formation and ongoing development of cities. Whereas urban planners and elected officials tend to think of sexuality as an exclusively private affair, sexual norms – namely, an expectation of monogamous, reproductive heterosexuality – serve as a foundation and organizing principle of both social relations and capitalist urban development.

Yet, sexuality does not operate on its own. Understandings and interpretations of sexual and gender diversity are co-constructed with notions of race, class, and other markers of social difference (Capó 2017; Somerville 2000). They are likewise central to the constitution of the settler-colonial nation-state: the

DOI: 10.4324/9781003157588-4

enforcement of gender-normative hetero-patriarchy has been and continues to be a defining feature of social relations (Dorries and Harjo 2020; Hunt 2016). These concepts are also context-dependent: time and place matter to how sexual and gender diverse lives are understood.

The acronym LGBTQ+ is a good example: a convention of late 20th century North America, the acronym stands for lesbian, gay, bisexual, transgender, and queer, where the plus sign denotes the many who are unnamed by the acronym but who may fit under the umbrella of sexual and gender diverse lives (including Two-Spirit, transsexual, intersex, asexual, pansexual, and questioning). The acronym, including some of the terms within it, does not easily translate across culture and language. As important, a diversity of LGBTQ+ experiences across race, class, and so on means there is a diversity of LGBTQ+ needs. Yet, many of the political and social gains and losses experienced by LGBTQ+ people bind them together under an internally diverse and often uneasy umbrella. The common denominator is the way that sexual and gender diverse people challenge the hetero-norms and expectations of society and the state.

This chapter provides a sketch of LGBTQ+ lives in North American cities since the turn of the 20th century. It is neither a global overview nor should it be read as a universal narrative: even between the United States and Canada, federal legislation and social welfare provision make for distinct queer histories. Still, these stories offer an important window on the shifting relationship between the state and LGBTQ+ communities, who continue to be perceived as deviant where their diversity is understood as an affront to the normative values espoused by the state.

1900–1945: the modern sexual subject under development

The visibility and shape of sexual and gender diverse landscapes in the early- to mid-20th century differ dramatically depending on where one looks. There were some common influences: progressive era reforms, federal New Deal policies (Canaday 2009), the advent of medicalized understandings of sexuality (Duder 2011), and the role that immigration and labor played across local economies. How these influences played out locally is a different story.

According to George Chauncey (1994), gay life in early 20th-century New York was visible and integrated into the mainstream. There was comparatively easy access to queer sexual and gender practices within its two established centers of queer life, Harlem and Greenwich Village; Coney Island – on the beach and under the boardwalk – was another site where queer sex was enjoyed (Hartman 2019, 313–315).

Some regional histories tell similar stories. In Minnesota, the homosocial character of extractive resource industries was praised as productive, even when gender-transgressive acts like cross-dressing were documented by industry inspectors (Murphy and Urquhart 2010, 45–49). Archival newspaper evidence suggests that although sex between men attracted some attention in the scandals of the day, alcohol was the greater concern and anti-sodomy laws, though exist-ent, were not enforced (Murphy and Urquhart 2010). Korinek (2018, 29) shows that men in Winnipeg in the 1930s accessed a public yet under-the-radar scene of cruising and drag by being attentive to suggestive hints: "a code word, flam-boyant clothes, make-up, or teenagers and young men engaged in 'swishy' or 'fairy-like' gender transgressive behaviours."

Yet, the story was altogether different elsewhere. In Portland, Oregon, two scandals related to men's same-sex practices in 1912–1913 were received with shock; social reformers used these scandals to toughen existing laws and create new regulations. The effects of these reforms lingered for at least 50 years and influenced politics and legal developments across the Pacific Northwest region (Boag 2003). There was equally swift action against the emergence of the United States' first gay rights group in Chicago in 1924, where the founder was arrested and lost his job with the U.S. Post Office for inappropriate conduct (Vaid 1995, 40). In Miami, white officials used charges of transgressive sexuality and gender performance to control and criminalize racialized communities, which in turn provided the city with coerced labor (Capó 2017).

Visibility of same-sex attraction carried different consequences for different groups. For middle- and upper-class white women, who were bound by con-ventions of respectability and had less access to public space, discretion meant keeping safe the reputations of themselves and their families (Duder 2011). For prominent Black women performers like Gladys Bentley and Jackie Mabley, success seemed to protect them from criticism for wearing men's clothes or loving women openly (Hartman 2019, 337). This has a geographic component as well: a relative freedom and anonymity was afforded in cities (Aldrich 2004), but queerness existed, if more quietly, in small towns as well (Duggan 1993; Howard 1999).

The meaning and significance attached to same-sex practices and non-normative gender performance underwent a dramatic shift during this period. Same-sex sexual acts and desire did not determine one's sense of self as gay or lesbian (Chauncey 1994; Duggan 1993). Neither "men like that" nor "men who like that" identified themselves in sexual terms (Howard 1999). The develop-ment of an identity based on one's non-normative sexual and gender practices and its coalescence into a (relatively) unified concept (Duder 2011) depended on multiple factors: medicalization of sexual deviance and gender nonconformity, reforms aimed at social control, and local urban "clean-up" campaigns that con-tributed to driving gay socializing underground.

1945–1965: the upsides and downsides of spatial segregation

One hallmark of the post-WWII period in North American cities is spatial seg-regation. Racial segregation was codified through lending and housing practices, and segregation of LGBTQ people resulted from both exclusion from mort-gage access (Howard 2013) and discrimination in housing and labor markets.

In San Francisco, trans women lived in the Tenderloin district because discrimination in housing and employment led them to a neighborhood where rent was cheap and difference was "tolerated" (Stryker and Silverman 2005). In Toronto, gays clustered in less attractive areas, with the upshot that they were able live more openly (Nash and Gorman-Murray 2015, 90).

Yet, the costs of being suspected or "found" frequenting a gay bar or sauna were tremendous. This era was marked by extreme homophobia founded on fears of sexual deviance (Brink 2012, 827). Homosexuality – which referred to same-sex attraction but also included diverging from gender norms – was considered a mental illness and was thus a basis for housing discrimination and termination of employment. Likewise, dressing in ways that did not conform to "proper" femininity for women or masculinity for men made people targets for harassment and arrest. Women found in gay bars were detainable for wearing pants with zippers in the front, according to Chicago's Municipal Code (Enke 2007, 50). The amplified attention to normative gender roles was a signature of the post-war and Cold War period (Boyd 2003). Women risked being read as "masculine" for not abandoning paid employment to which they gained access during the war effort (Chenier 2004).

Despite intimidation, people gathered and socialized. In San Francisco, the gay bar emerged as a new kind of establishment distinct from the mixed venues of the pre-war era. Unlike earlier meeting places, gay bars allowed for self-disclosure, which facilitated a new level of community development that included language and cultural practices. Bars began to use spatial defenses: back entrances, door management, and other mechanisms to ensure patrons' privacy, while also protecting them from police raids (Boyd 2003, 104). Race and class shaped who had access to bar culture (Chenier 2004; Lorde 1982). Gathering places were not always public, however: Black lesbians in Detroit (Thorpe 1996) and in Buffalo (Kennedy and Davis 1993) socialized in private spaces, including each other's homes (Nestle 1993, 9).

While there were many who participated in same-sex practices but did not think of themselves as gay or as part of a gay community (Stein 2014), a significant increase in a community consciousness occurred during this time. This was aided by gay networks, bar culture, and print media (newsletters, magazines, and novels). Homophile organizing, which sought to assimilate LGBTQ+ people into mainstream society, was also important. Homophile ideology focused on "individual civil rights based on the right to privacy rather than the right to public association" (Boyd 2003, 160). The conflict between homophile organizing and militant responses to police and social repression is a thread that weaves through all subsequent eras.

1965–1980: asserting identity and building community

The period of social activism and agitation over civil rights and sexual liberation was marked by protests and by the formation of gay villages, also called gayborhoods. While comparable territories existed earlier, it was not until after the sexual liberation of 1960s that certain neighborhoods both connoted and represented gay (male) sexuality (Brown 2014, 458). Still located in more marginal sections of cities (Hess and Bitterman 2021, 11) but now highly visible enclaves, gayborhoods offered a different level of gay identity and community development (Brown 2014). Visibility of neighborhoods meant that residents and users were also more visible, which translated to greater recognition in the public sphere. Indeed, electoral victories showed that neighborhoods were the basis for political gains. Neighborhoods were perceived "as an expressive demonstration of gay identity, and thus as the collective asset most in need of protection" (Hanhardt 2008, 67).

In the United States, the rise of self-organized street patrols was one form of protection and political organizing that occurred in gay neighborhoods (Hanhardt 2008). Patrols explicitly focused on calling out anti-gay violence and exploited racialized and classed discourses to identify those responsible for these actions (Hanhardt 2008). Another form was organizing for gay civil rights legislation, especially at the municipal level. In New York, the Gay Activist Alliance focused most of its attention on advocating for a gay civil rights bill (Hanhardt 2008), and in Minneapolis, the leader of the city's first gay political group filed for a marriage license with his lover (Knopp 1987).

These forms of organizing demonstrate a shift both in tactics and in imagining the target of LGBTQ+ activism. Activists in the 1970s shifted their attention away from militant protests against police violence; instead, they began to focus on violence carried out by random strangers (Hanhardt 2008). The shape of political organizing shifted toward "mainstreaming": a move away from efforts to radical social change that were associated with the Stonewall riots[1] and similar anti-state antagonism, toward a politics and a social movement that focused on legitimation (Vaid 1995).

This strategic shift should be understood in relation to the rise of anti-gay politics in the United States. In the 1970s, there was an uptick in gay bashings and murders, which was doubly threatening when the police remained a source of repression (Murphy and Urquhart 2010). Anti-gay political campaigns also picked up in 1977–1978 following the implementation of non-discrimination ordinances that included lesbians, gays, and bisexual people (Dubrow et al. 2015). Anita Bryant's "Save Our Children" led the way in what would become a long-term, multi-faceted, religiously-based campaign to legalize anti-gay policies (Dubrow et al. 2015). The organization led successful, funded campaigns to repeal non-discrimination ordinances in targeted cities while also championing initiatives aimed at preventing or removing gay people from public school teaching positions (Dubrow et al. 2015).

Political organizing in Canada remained militant in this period in response to police repression. As in the United States, there was increased visibility of gay and lesbian community-building. However, being gay in public, whether cruising and having public sex or being found in a bar or a bathhouse, contravened the Criminal Code and offered a rationale for raids on, and closures of, gay and lesbian establishments. Activists confronted the Criminal Code and police repression with protest (Kinsman 2017). The use of print media in the form of local newsletters and national magazines like *Body Politic* was also instrumental to organizing for change.

Three other points are significant. First, gay neighborhoods were not the only spaces to serve an important function in this period. Spaces for cruising and public sex were critical: sexual opportunities developed in tandem with the available urban infrastructure, such as abandoned piers and public parks, as well as with privately-run spaces like porn theatres (Escoffier 2017). In many cities, including those outside of large metropolitan centers, universities played an important role for community building. The University of Manitoba's Gays for Equality played a crucial role in Winnipeg: based at the university, the group fostered the development of community through events, campaigns, and organizations both on campus and in the city (Korinek 2018).

Second, this period saw a surge of new organizations, which contributed to the advancement of LGBTQ+ as an identity. Seventeen groups sprouted up in Canadian cities between 1969 and 1976 (Tremblay 2015, 16–17), but there were surely others: a 1977 conference held in Winnipeg brought together eleven organizations from across Manitoba alone (Korinek 2018). Organizations had various missions and functions, from political activism and education to coordinating social events and providing peer support.

Finally, many women carved out their own spaces during this period. Lesbians developed communal living spaces: this included rural communal houses, cooperative living, and taking over large houses in less expensive parts of urban centers (Millward 2015, 122–123). Some of these were separatist spaces; some were "women-identified," which was a political designation rather than a sexual one (Chamberland 2019, 187). Others were living collectives; Pride and Prejudice in Chicago combined collective housing with a bookstore and a community center (Enke 2007). While Pride and Prejudice was a feminist collective and not exclusively lesbian, it was essential to the lesbian scene in Chicago, co-producing the first all-women's dance and lesbian feminist band in the city (Enke 2007).

Gathering spaces were just as critical. These included resource centers, coffee houses, and softball pitches (Enke 2007). Few were able to raise enough money to buy a building, as was the case with the Women's Building in Winnipeg. However, many groups were able to establish themselves by holding regular events and meetings in these locations and thus becoming associated with these spaces (Millward 2015, 125). Recurring event spaces, which could be found in parks, churches, and businesses, were as important as unique, short-term gatherings like conferences (Nusser and Anacker 2013). According to Yolanda Retter (1997, 332), 1,500 lesbians convened for the West Coast Lesbian Conference held in Los Angeles in 1973.

1980–2000: from marches to parades

The year 1981 marked three significant moments for North American LGBTQ+ communities that informed the two decades leading to the new millennium and beyond. In 1981: (1) The U.S. Centers for Disease Control identified the first AIDS case. (2) The Toronto bathhouse raids resulted in the arrest of nearly three hundred people; this event gave rise to massive protests in Toronto, which subsequently became annual Pride march (Bain 2017, 82). And (3) in Vancouver, the first Gay Unity Week was proclaimed by City Council in recognition of the contributions of gays and lesbians to the city (Murray 2015/16, 58). Each phenomenon offers a window into these two decades.

In the United States, the effect of the AIDS epidemic in its height was immense and unrelenting. Peaking between 1987 and 1996, AIDS deaths accounted for one in four deaths of men aged 25–44 in 1995 (CDC 2011, cited in Rosenfeld et al. 2012). Yet, deaths were unevenly distributed: AIDS was a distinctly urban phenomenon, where it wreaked havoc on the gay neighborhoods and networks that had come up in the previous twenty years (Rosenfeld et al. 2012). Moreover, the death rate was higher among Black men, and women initially were undercounted because their symptoms were not included in the definition of the virus.

One effect of the crisis was that gay villages became even more critical to LGBTQ+ life as the services provided – counseling, meal programs, hospice, and more – were a community response not offered elsewhere (Nash and Gorman-Murray 2015, 87). Another effect was the rapid pace of neighborhood change where gay men died. AIDS-related deaths and the displacement of lovers who did not have legal protections or recourse to keep their housing resulted in the conversion of New York's West Village (and beyond) from racially diverse and mixed income to "market rate" and homogenously wealthy (Schulman 2012).

The emergence of ACT UP, the direct-action activist movement that emerged to respond to the AIDS crisis, was another outcome. ACT UP organized highly visible actions to confront homophobia, racism, and sexism in the treatment of People With AIDS and the inaction of public health officials and pharmaceutical companies, among others (Schulman 2021). The tactics deployed by ACT UP – its reliance on disruptive, theatrical protest and use of graphic design and media – show a trajectory of civil rights disobedience that continues to shape social activism (Shephard and Hayduk 2002).

Sexual policing persisted amid some legal gains. In the 1980s, it was common for LGBTQ+ people to be refused service at restaurants (Schulman 2012, 29); be omitted from the definition of "family housing" (Forsyth 2011); be censored (Zanin 2017); and be excluded from a lover's hospital room, health care decisions, and social welfare benefits (Valenti and Katz 2014). Yet, between 1982 and 1995, gay rights ordinances were passed in nearly 150 U.S. cities and counties and nine U.S. states (Vaid 1995). In Canada, as in the United States, many municipalities enacted far more progressive social policies than federal governments or even provincial human rights legislation. Vancouver's ban on gender discrimination in its human rights policy encompassed transgender people well before this was incorporated in the provincial human rights code (Murray 2015/16).

Any gains made were highly uneven, however, and there was persistent backlash by conservative groups (Millward 2015). Anti-gay ballot measures became commonplace in the U.S; nearly forty passed between 1991 and 1995 (Vaid 1995).

Likewise, police raids of LGBTQ+ spaces were unexceptional through the mid-1980s in Ontario (Rayside 2015). In fact, the last notable Toronto police raid was conducted in 2000 at Pussy Palace, a queer women's bathhouse event held as part of Pride. In Montreal, the violence used by police during a raid on a warehouse party in 1990 sparked a sit-in and kiss-in attended by several hundred during the following two days (Podmore 2015).

Occupying public space was a key response to repression. Protests in response to the 1981 Toronto bathhouse raids involved thousands and took place over the next two years (Grube 1997). Demonstrations against police harassment and policies harmful to lesbians gave rise to the first two lesbian pride marches in North America in 1981 (Burgess 2017). The Dyke March, initiated by Lesbian Avengers as part of the 1993 March on Washington, grew out of this tradition, demanding space for women and visibility for lesbians without seeking permits or authorization (Burgess 2017).

Tactics changed during this period. Responding to gay bashing meant cultivating better relations with police (Podmore 2015). The promotion of certain neighborhoods and collaboration with urban clean-up campaigns meant the active displacement of sex workers (Ross and Sullivan 2012). The shape of Pride events was debated. ACT-UP and Queer Nation Rose protested the Montreal Pride parade twice in the early 1990s: first, to take the march to a more visible location rather than stick to the gay village, and again to resist a dress code created by local organizers (Bell and Valentine 1995). Likewise, Toronto Pride underwent a commercial transformation. Whereas the organizing of Pride was initiated in 1986 to make the event accessible to whomever wanted to join, Toronto Pride became a party atmosphere that excluded Black and racialized queers (Bain 2017). While Kath Browne (2007) theorizes Pride as "parties with politics," the transformation from protest march to celebratory parade is suggestive of the changing landscape for LGBTQ+ lives (Hughes 2003, cited in Browne 2007).

2000–present: queers and the state make uneasy bedfellows

The turn of the 21st century saw a greater formalization in the relationship between the mainstream lesbian and gay movement and the state. In Canada, federal changes in 2000 gave same-sex couples access to rights and benefits (Tremblay 2015). Subsequent actions in provincial and territorial courts brought civil marriage by 2004. Equal marriage was understood as a victory, but not all supported a singular focus on this policy issue (Tremblay 2015, 22). Some opposed marriage itself, whether because of its patriarchal history or because of the limited range of (well-resourced) people it serves. Some resisted the fact that one issue absorbed so much energy at the expense of many other concerns.

The conflict over marriage became the prime example of the long-standing tension between those who focus on civil rights-based gains and those who seek broader structural change beyond the state. This tension was crystallized in the term homonormativity, which posits gay culture as assimilationist, depo-liticized, and focused on the trappings of coupledom (Duggan 2002). Debates over the politics of homonormativity have served as the backdrop of this era: they informed the shape of LGBTQ+ efforts to work with the state to address community needs: for instance, calling upon municipalities to underwrite the development of dedicated gathering space or to protect senior housing. These debates also inform who is called upon, and who feels entitled, to participate in community visioning and planning practices.

Four themes are critical in this period: the ongoing effects of previous poli-cies; the new perception of LGBTQ+-themed events as an economic boon for cities; the status of gay villages; and the adoption of some LGBTQ+-related content in City Hall.

1. The effects of persistent segregation. The legacy of racially segregated land-scapes continues to play a defining role in the shape of LGBTQ+ communities. The same political, economic, and sociocultural factors that normalized racially divided cities in North America restricted queers of color from participating in LGBTQ+ spaces in neighborhoods other than their own (Greene 2019). There is thus a persistent failure of the dominant movement to recognize LGBTQ+ diversity. This misrecognition has knock-on effects for both queers of color and those who fit within the dominant mold (Muñoz 1999; Nash 2013). The rise of everyday violence directed especially at queers of color in the post-9/11 period illustrates these effects. Practices like dispensing with public-sex spaces, which had previously been framed as "revitalization" began to be framed in terms of "national security" (Manalansan 2005, 146). The effects were chilling for com-munities of color who were already dealing with the impacts of gentrification and displacement. At the same time, these neighborhoods were being advertised in mainstream (white) gay press as exotic spaces for (white) visitors' consumption (Manalansan 2005).

2. The economic benefit from LGBTQ+ events. As North American cities increas-ingly rely on leisure and tourism as elements of entrepreneurial and creative strat-egies, LGBTQ+-themed events are understood as economic drivers for cities and offer a promise of cashing in on "pink tourism." In 2014, Toronto hosted World Pride; the economic impact of its two million attendees was reported to be CDN\$719 million (Armstrong 2015). In Miami Beach, where two annual events target Latinx LGBTQ+ women and people of color, pink tourism is highly lucra-tive. According to the President of the Miami-Dade Gay and Lesbian Chamber of Commerce, pink tourism is estimated to bring in US\$1.7 billion annually to the local economy (Squires 2019). The financial stakes of these events have resulted in greater support from cities, but only to comparatively well-resourced LGBTQ+ groups and events (McLean 2014). Reliance on municipal (and cor-porate) support also requires complying with external demands. In 2010, the Toronto Pride Committee was threatened with losing municipal funding over the participation of Queers Against Israeli Apartheid, an activist group (Bain 2017). In response, Toronto Pride prevented the group from marching in the

Pride parade, setting off a storm of protests over censorship and homonormative depoliticization (Bain 2017).

3. *Gay villages*. The gay village – its relevance and demise – has been in question (Ghaziani 2014; 2021). Toronto's village has experienced increased rents and growth of market-rate housing; this has meant the shuttering of symbolic stores and changing neighborhood demographics (Nash and Gorman-Murray 2015, 92). Should the affordability and unique qualities of gay villages be preserved (Doan 2015)? If so, what role should municipalities play? Gay villages are symbolic but are also understood as sex-positive places where public intimacy is possible, in contrast to suburban locales (Brink 2012). They feature LGBTQ+-serving and -affirming businesses, greater access to sex (bathhouses, cruising spaces), and an expected freedom from homophobic surveillance. Moreover, gay villages offer visible recognition of the LGBTQ+ community, even where historical antecedents have been absent. In Ottawa, Le/The Village achieved a municipal designation in 2011 and provided visibility in a city with an extensive history of repressing LGBTQ+ lives (Lewis 2013). However, gay villages are exclusionary and reflect the dominant narrative of gay visibility: white, male, cisgender,[2] and affluent. While Greene (2019) argues that gayborhoods are perceived to be spaces where a gay couple can kiss, he also denotes how practices employed by queer youths of color are read by some (white) residents as a threat to public safety. For Black youth who try to carve out space during Pride, the Toronto village is a space of unbelonging where their bodies and their needs are repeatedly de-prioritized (Rosenberg 2021). Exclusions can be grounded in commercial interests. In Montreal's Gay Village, the municipal and merchants responded to the presence of homeless people by using fear of anti-gay violence to guard against what property owners consider to be unseemly street life (Podmore 2015, 201). Myriad factors inform how queers use or perceive the gayborhood. Life course is one consideration: migration to and from gay districts at different points in life is influenced in part by a desire to be in proximity to gay men (Lewis 2014). Attachment to place is another factor. Some claim a sense of identity from residing in, or being close to, a queer-identified neighborhood (Doan and Higgins 2011). Suburbanization plays another significant role in the use and perception of gay villages (Anacker 2011). In Canadian cities, the colossal disparity between housing prices and incomes has pushed many to seek housing and community beyond the urban core. In municipalities outside of Vancouver, LGBTQ+ activism takes a variety of forms, from peer support for LBGTQ+ South Asian young adults to annual Pride celebrations and municipal proclamations (Bain and Podmore 2021). Activisms are uneven across suburban geographies, but they indicate the making of safe spaces beyond the gayborhood. Despite concerns, many reject the notion that gay villages are on their way out. Instead, multiple neighborhoods are understood to sustain LGBTQ+ communities as they cultivate safe spaces (Nash and Gorman-Murray 2015, 92). Ghaziani (2019, 12) describes a cultural archipelago: rather than a singular LGBTQ+ district, several territories exist and function in relation to and in tension with

one another. Greene (2019) suggests that *queer* cultural archipelagos offer a better representation of the spatial complexity of LGBTQ+ communities and their vitality, which, he argues, have always extended beyond the confines of one neighborhood. Municipal planning also plays an active role in retaining (or not) distinctive gay neighborhoods. Municipal (in)decisions can make a difference in an area's retention of gay-related content, as Doan and Higgins (2011) show in Atlanta. Here, efforts to stimulate neighborhood redevelopment in some areas worked in conjunction with disregard of other areas, alongside uneven application of policies like liquor licensing (Doan 2015). These tactics drove gay institutions away from a central historic neighborhood toward peripheral areas. The Atlanta case study also showed a notable absence of the diversity of LGBTQ+ voices in the planning process (Doan and Higgins 2011).

4. *Queer life in City Hall*. A final theme, especially relevant since 2010, is that LGBTQ+ concerns are increasingly recognized within City Hall. In some communities, these are one-off or short-term engagements: approving a petition to raise the Pride flag, for instance, or painting a temporary rainbow crosswalk (see Case study). In others, citizen/resident advisory committees have been established to offer guidance, feedback, or advocacy on city policy or strategy (Murray 2015/2016). Hamilton and St. Catharines (Ontario) and Vancouver all have LGBTQ+ Advisories. In Vancouver, these had a tremendous effect. Work undertaken in 2014 by an LGBTQ+ advisory sub-committee for the Vancouver Board of Parks and Recreation produced a report identifying seventy-seven recommendations to make park and recreation services accessible to transgender and gender diverse people (TGVI 2014). This report fostered immediate changes in service provision and programming, accompanied by a highly visible educational campaign targeting all park users, staff training, and changes to washroom signage and design.

Conclusion

Much has changed for LGBTQ+ communities in North America: there is no fear of being arrested or having one's name printed in the newspaper for being found in a gay bar. Same-sex relationships are recognized by the state and many faith communities. LGBTQ+ lives are part of mainstream culture. Yet, subtle and overt discrimination remain, and forces aiming to deny legitimacy and human rights to some or all of those under the LGBTQ+ umbrella are never distant. Bathroom laws and policing of sport participation are two examples of recent attacks that have especially targeted transgender and gender diverse people. Internal community conflicts, stemming from generational differences (Nash 2013; Podmore 2021) and racism (Bain 2017), also pose a challenge.

Planners have the opportunity and the mandate to assess whose needs are being met by existing municipal plans and service provision. In many municipalities, the inclusion of LGBTQ+ communities is very recent and exceptional and, as Bain and Podmore (2020; 2021) show, unevenly present within broader commitments to equity and social inclusion. However, LGBTQ+ perspectives are critical to shaping urban futures: LGBTQ+ communities bring specific needs to concerns like aging in place and the safety and design of public spaces. Likewise, inclusion must attend to more than sexual diversity in order to capture the kinds of experiences that queer and transgender people of color have when using city services. By drawing from LGBTQ+ expertise, planners can fill the gaps that were created when cities were planned for an imagined universal user. Doing so will bring forth a more inclusive city.

Alternative planning history and theory timeline

Period	Planning history and theory canon	Alternative timeline: LGBTQ+ communities
1900–1945	Birth/formalization of planning	*Development of modern sexual subject* Visible gay life in New York and some major cities, "clean up" campaigns and regulations elsewhere, nascent gay or lesbian identity
1945–1965	Growth of planning	*Spatial segregation* Exclusion from mortgage access, discrimination in housing and labor markets, emerging gay bar scene, emerging community consciousness
1965–1980	Midlife crisis of planning	*Asserting identity and building community* Social activism, sexual liberation, gay bashing and murders, gay villages, lesbian communes, spaces for cruising and public sex, advancement of LGBTQ+ identity
1980–2000	Maturation of planning	*From marches to parades* AIDS epidemic, civil rights disobedience, anti-gay ballot measures, visible marches and parades
2010– present	New planning crisis	*Queers and the state* Same-sex couples access to rights and benefits, homonormativity, persistent segregation, LGBTQ+ events, Pride flags, rainbow crosswalks

Case study: the rainbow crosswalk

The rainbow crosswalk is a common feature of the urban landscape in Canadian towns and cities. Located at symbolic intersections – in the gay village or in front of city hall – or in high visibility areas, the rainbow crosswalk is a pedestrian crossing painted in an "eight colour rainbow scheme [that] reflects the original Pride flag colours from 1978, symbolizing diversity and inclusivity" (City of Vancouver 2013, 16). Vancouver claims to have established the first permanent rainbow crosswalks in Canada in 2013. Municipalities of all sizes have followed suit: from Maskwacis, a community of Cree First Nations near Red Deer (Alberta) and Wolfville (Nova Scotia), to Moncton (New Brunswick) and Montreal. There is a diversity of opinion about the purpose of the rainbow crosswalk, but it is widely understood as a symbol of an inclusive community that embraces LGBTQ+ residents and social diversity more broadly. For many, the rainbow crosswalk symbolizes acceptance. One city councilor reflected: "As a closeted gay kid growing up in Barrie [Ontario], a symbol like that would have gone a long way toward letting me know that if you're a member of the LGBTQ community, you have a place here in Barrie" (as reported in Owen 2019). It is also a policy accompanied by controversy. Sometimes, this is couched within other official protocols, like traffic safety or existing bylaws. For example, many municipalities in New Brunswick cite an in-progress Transportation Association of Canada study on Non-Standard Pavement Markings for Crosswalks when determining whether to create rainbow crosswalks. More often, it is framed as a political – and thus potentially divisive – statement. Officials in Chilliwack (British Columbia) cited this reasoning when the City Council rejected the proposal for one downtown rainbow crosswalk in 2019. In response, community members painted their own rainbow crosswalks. By the end of that year, there were 16 rainbow crosswalks completed or in development: across private driveways and otherwise on residential properties, on First Nations-governed land, and on public school properties (Woods 2019). Municipalities should use public space to support equity and inclusion, and rainbow crosswalks are one strategy in the planning toolbox. Yet, creating visibility for LGBTQ+ communities is meaningful only when it is accompanied by commitments that advance LGBTQ+ human rights.

Notes

1 The Stonewall riots were a series of demonstrations in 1969 by members of the LGBTQ+ community in response to a police raid at the Stonewall Inn in New York City.
2 Cisgender means that there is alignment between anatomy and gender identity.

References

Aldrich, Robert. 2004. "Homosexuality and the City." *Urban Studies* 41 (9): 1719–1737.

Anacker, Katrin. 2011. "Queering the Suburbs." In *Queerying Planning*, edited by Petra Doan, 107–125. Surrey: Ashgate.

Armstrong, Laura. 2015. "Who Wins the Battle of the Toronto Summer Events?" *Toronto Star*, August 21. https://www.thestar.com/news/gta/2015/08/21/who-wins-the-battle-of-the-toronto-summer-events.html.

Bain, Beverly. 2017. "Fire, Passion, and Politics." In *We Still Demand!*, edited by Patrizia Gentile, Gary Kinsman, and L. Pauline Rankin, 81–97. Vancouver: UBC Press.

Bain, Alison, and Julie Podmore. 2020. "Scavenging for LGBTQ2S Public Library Visibility on Vancouver's Periphery." *Tijdschrift voor economische en sociale geografie* 111 (4): 601–615. https://doi.org/10.1111/tesg.12396.

Bain, Alison, and Julie Podmore. 2021. Linguistic ambivalence amidst suburban diversity: LGBTQ2S municipal 'social inclusions' on Vancouver's periphery. EPC: Politics and Space 39 (7): 1644–1672. https://doi.org/10.1177/23996544211036470.

Bell, David and Gill Valentine. 1995. "Introduction: Orientations." In *Mapping Desire*, edited by David Bell and Gill Valentine, 1–27. New York: Routledge.

Boag, Peter. 2003. *Same-Sex Affairs*. Berkeley: University of California Press.

Boyd, Nan Alamilla. 2003. *Wide Open Town*. Berkeley: University of California Press.

Brink, Charles J. Ten. 2012. "Gayborhoods: Intersections of Land Use Regulation, Sexual Minorities, and the Creative Class." *Georgia State University Law Review* 28 (3): 789–849.

Brown, Michael. 2014. "Gender and Sexuality II: There Goes the Gayborhood?" *Progress in Human Geography* 38 (3): 457–465.

Browne, Kath. 2007. "A Party with Politics?" *Social & Cultural Geography* 8 (1): 63–87.

Burgess, Allison. 2017. "The Emergence of the Toronto Dyke March." In *In We Still Demand!*, edited by Patrizia Gentile, Gary Kinsman, and L. Pauline Rankin, 98–116. Vancouver: UBC Press.

Canaday, Margot. 2009. *The Straight State*. Princeton: Princeton University Press.

Capó, Julian. 2017. *Welcome to Fairyland: Queer Miami Before 1940*. Chapel Hill: University of North Carolina Press.

Centers for Disease Control (CDC). 2011. Mortality slide series. Retrieved May 8, 2011 from http://www.cdc.gov/hiv/topics/surveillance/resources/slides/mortality/slides/mortality.pdf.

Chamberland, Line. 2019. "The Place of Lesbians in the Women's Movement." In *Lesbian Feminism*, edited by Niharika Banerjea, Kath Browne, Eduarda Ferreira, Marta Olasik, and Julie Podmore, 184–201. London: Zed Books.

Chauncey, George. 1994. *Gay New York: Gender, Urban Culture, and the Making of the Gay Male World, 1890–1940*. New York: Basic Books.

Chenier, E. 2004. "Rethinking Class in Lesbian Bar Culture: Living 'The Gay Life' in Toronto, 1955–1965." *Left History Review* 9 (2): 85–118. https://doi.org/10.25071/1913-9632.5608.

City of Vancouver. November 2013. "West End Community Plan." https://vancouver.ca/files/cov/west-end-community-plan-2013-nov.pdf.

Doan, Petra. 2015. "Why Plan for the LGBTQ Community?" In *Planning and LGBT Communities*, edited by Petra Doan, 1–15. New York: Routledge.

Doan, Petra, and Harrison Higgins. 2011. "The Demise of Queer Space?" *Journal of Planning Education and Research* 31 (6): 6–25.

Dorries, Heather, and Laura Harjo. 2020. "Beyond Safety: Refusing Colonial Violence through Indigenous Feminist Planning." *Journal of Planning Education and Research* 40 (2): 210–219.

Dubrow, Gail, Michael Brown, and Larry Knopp. 2015. "Act Up versus Straighten Up." In *Planning and LGBT Communities*, edited by Petra Doan, 202–216. New York: Routledge.

Duder, Cameron. 2011. *Awfully Devoted Women*. Vancouver: University of British Columbia Press.

Duggan, Lisa. 1993. "The Trials of Alice Mitchell." *Signs* 18 (4): 791–814.

Duggan, Lisa. 2002. "The New Homonormativity." In *Materializing Democracy: Toward a Revitalized Cultural Politics*, edited by Russ Castronovo and Dana Nelson, 175–194. Durham: Duke University Press.

Enke, A. 2007. *Finding the Movement*. Durham: Duke University Press.

Escoffier, Jeffrey. 2017. "Sex in the Seventies: Gay Porn Cinema as an Archive for the History of American Sexuality." *Journal of the History of Sexuality* 26 (1): 88–113.

Forsyth, Ann. 2011. "Queerying Planning Practice." In *Queerying Planning*, edited by Petra Doan, 21–51. Surrey: Ashgate.

Ghaziani, Amin. 2014. *There Goes the Gayborhood?* Princeton: Princeton University Press.

Ghaziani, Amin. 2019. "Cultural Archipelagos: New Directions in the Study of Sexuality and Space." *City & Community* 18 (1): 4–22. https://doi.org/10.1111/cico.12381.

Ghaziani, Amin. 2021. "Why Gayborhoods Matter: The Street Empirics of Urban Sexualities." In *The Life and Afterlife of Gay Neighbourhoods: Renaissance and Resurgence*, edited by Alex Bitterman and Daniel Baldwin Hess, 87–113. Cham, Switzerland: Springer https://doi.org/10.1007/978-3-030-66073-4.

Greene, Theodore. 2019. "Queer Cultural Archipelagos are New to Us." *City & Community* 18 (1): 23–29.

Grube, John. 1997. "'No More Shit': The Struggle for Democratic Gay Space in Toronto." In *Queers in Space*, edited by Gordon Brent Ingram, Anne-Marie Bouthillette, and Yolanda Retter, 127–145. Seattle: Bay Press.

Hanhardt, Christina. 2008. "Butterflies, Whistles, and Fists: Gay Safe Streets Patrols and the New Gay Ghetto, 1976–1981." *Radical History Review* 100: 61–85.

Hartman, Saidiya. 2019. *Wayward Lives, Beautiful Experiments*. New York: Norton.

Hess, Daniel Baldwin, and Alex Bitterman. 2021. "Who Are the People in Your Gayborhood?" *In The Life and Afterlife of Gay Neighbourhoods: Renaissance and Resurgence*, edited by Alex Bitterman and Daniel Baldwin Hess, 3–39. Cham, Switzerland: Springer https://doi.org/10.1007/978-3-030-66073-4.

Howard, Clayton. 2013. "Building a 'Family-Friendly' Metropolis: Sexuality, the State, and Postwar Housing Policy." *Journal of Urban History* 39 (5): 933–955. https://doi.org/10.1177/0096144213479322.

Howard, John. 1999. *Men Like That*. Chicago: University of Chicago Press.

Hubbard, Phil. 2001. "Sex Zones: Intimacy, Citizenship, and Public Space." *Sexualities* 4 (1): 51–71. https://doi.org/10.1177/136346001004001003.

Hunt, Sarah. 2016. *An Introduction to the Health of Two-Spirit People: Historical, Contemporary and Emergent Issues*. Prince George, BC: National Collaborating Centre for Aboriginal Health. www.ccnsa-nccah.ca/docs/emerging/RPT-HealthTwoSpirit-Hunt-EN.pdf.

Kennedy, Elizabeth Lapovsky, and Madeline D. Davis. 1993. *Boots of Leather, Slippers of Gold*. New York: Routledge.

Kinsman, Gary. 2017. Queer Resistance and Regulation in the 1970s. In *We Still Demand!*, edited by Patrizia Gentile, Gary Kinsman, and L. Pauline Rankin, 137–162. Vancouver: UBC Press.

Knopp, Lawrence. 1987. "Social Theory, Social Movements and Public Policy." *International Journal of Urban and Regional Research* 11 (2): 243–261.

Korinek, Valerie. 2018. *Prairie Fairies*. Toronto: University of Toronto Press.

Lewis, Nathaniel. 2013. "Le/The Village: Creating a 'Gaybourhood' Amidst the Death of the Village." *Geoforum* 49: 233–242. https://doi.org/10.1016/j.geoforum.2013.01.004.

Lewis, Nathaniel. 2014. "Moving 'Out', Moving On: Gay Men's Migrations Through the Life Course." *Annals of the Association of American Geographers* 104 (2): 225–233. https://doi.org/10.1080/00045608.2013.873325.

Lorde, Audre. 1982. *Zami: A New Spelling of My Name*. New York: Crossing Press.

Manalansan, Martin. 2005. "Race, Violence, and Neoliberal Spatial Politics in the Global City." *Social Text* 23 (3–4): 141–155.

McLean, Heather. 2014. "Digging into the Creative City: A Feminist Critique." *Antipode* 46 (3): 669–690. https://doi.org/10.1111/anti.12078.

Millward, Liz. 2015. *Making a Scene*. Vancouver: UBC Press.

Muñoz, José Esteban. 1999. *Disidentifications*. Minneapolis: University of Minnesota Press.

Murphy, Ryan Patrick, and Alex Urquhart. 2010. "Sexuality in the Headlines." In *Queer Twin Cities*, edited by Twin Cities GLBT Oral History Project, 40–89. Minneapolis: University of Minnesota Press.

Murray, Catherine. 2015/2016. "Queering Vancouver: The Work of the LGBTQ Civic Advisory Committee, 2009–14." *BC Studies* 188 (Winter): 55–80.

Nash, Catherine. 2013. "The Age of the 'Post-Mo'? Toronto's Gay Village and a New Generation." *Geoforum* 49: 243–252.

Nash, Catherine, and Andrew Gorman-Murray 2015. "Recovering the Gay Village: A Comparative Historical Geography of Urban Change and Planning in Toronto and Sydney." *Historical Geography* 43: 84–105.

Nestle, Joan. 1993. "Excerpts from the Oral History of Mabel Hampton." *Signs* 18 (4): 925–935.

Nusser, Sarah, and Katrin Anacker. 2013. "What Sexuality Is this Place?" *Journal of Urban Affairs* 35 (2): 173–193.

Owen, Jessica. 2019. "Rainbow crosswalk provides ray of hope to LGBTQ newcomers," *Bradford Today*, July 4, 2019. https://www.bradfordtoday.ca/local-news/rainbow-crosswalk-provides-ray-of-hope-to-lgbtq-newcomers-1554666.

Podmore, Julie. 2015. "From Contestation to Incorporation." In *Queer Mobilizations*, edited by Manon Tremblay, 187–207. Vancouver: UBC Press.

Podmore, Julie. 2021. "Far Beyond the Gay Village." In *The Life and Afterlife of Gay Neighbourhoods: Renaissance and Resurgence*, edited by Alex Bitterman and Daniel Baldwin Hess, 289–306. Cham, Switzerland: Springer https://doi.org/10.1007/978-3-030-66073-4.

Rayside, David. 2015. "Queer Advocacy in Ontario." In *Queer Mobilizations*, edited by Manon Tremblay, 85–105. Vancouver: UBC Press.

Retter, Yolanda. 1997. "Lesbian Spaces in Los Angeles, 1970–90." In *Queers in Space*, edited by Gordon Brent Ingram, Anne-Marie Bouthillette, and Yolanda Retter, 325–337. Seattle: Bay Press.

Rosenberg, Rae. 2021. "Negotiating Racialised (un)belonging: Black LGBTQ Resistance in Toronto's Gay Village." *Urban Studies* 58 (7): 1397–1413. https://doi.org/10.1177/0042098020914857.

Rosenfeld, Dana, Bernadette Bartlam, and Ruth D. Smith. 2012. "Out of the Closet and Into the Trenches." *The Gerontologist* 52 (2): 255–264. https://doi.org/10.1093/geront/gnr138.

Ross, Becki, and Rachael Sullivan. 2012. "Tracing Lines of Horizontal Hostility." *Sexualities* 15 (5/6): 604–621. https://doi.org/10.1177/1363460712446121.

Schulman, Sarah. 2012. *Gentrification of the Mind*. Berkeley: University of California Press.

Schulman, Sarah. 2021. *Let the Record Show: A Political History of ACT UP New York, 1987–1993*. New York: Farrar, Straus and Giroux.

Shephard, Benjamin and Ronald Hayduk. 2002. *From ACT UP to the WTO*. London: Verso.

Somerville, Siobhan. 2000. *Queering the Color Line*. Durham: Duke University Press.

Squires, Kai Kenttamaa. 2019. "Rethinking the Homonormative? Lesbian and Hispanic Pride Events and the Uneven Geographies of Commoditized Identities." *Social & Cultural Geography* 20 (3): 367–386. https://doi.org/10.1080/14649365.2017.1362584.

Stein, Marc. 2014. "Canonizing Homophile Sexual Respectability." *Radical History Review* 120: 52–73. https://doi.org/10.1215/01636545-2703724.

Stryker, Susan, and Victor Silverman. 2005. *Screaming Queens: The Riot at Compton's Cafeteria*. Frameline distributors. www.youtube.com/watch?v=G-WASW9dRBU.

Tallbear, Kim. 2019. "Decolonizing Sex." Produced by Matike Wilbur and Adrienne Keene. *All My Relations*, March 19, 2019. Podcast, 43:24. www.allmyrelationspodcast. com/podcast/episode/468a0a6b/ep-5-decolonizing-sex.

Thorpe, Rochella. 1996. "'A House Where Queers Go': African-American Lesbian Nightlife in Detroit, 1940–1975." In *Inventing Lesbian Cultures in America*, edited by Ellen Lewin, 40–61. Boston: Beacon Press.

Trans and Gender Variant Inclusion (TGVI) Working Group. 2014. "Building a Path to Parks and Recreation for All." *City of Vancouver*. April. Retrieved from https:// vancouver.ca/files/cov/REPORT-TGVIWorkingGroupReport-2014-04-28.pdf.

Tremblay, Manon. 2015. "Introduction." In *Queer Mobilizations*, edited by Manon Tremblay, 3–41. Vancouver: UBC Press.

Vaid, Urvashi. 1995. *Virtual Equality*. New York: Anchor Books.

Valenti, Korijna, and Anne Katz. 2014. "Needs and Perceptions of LGBTQ Caregivers." *Journal of Gay & Lesbian Social Services* 26: 70–90.

Woods, Mel. 2019. "Chilliwack, B.C. Becomes Unlikely Hub for Rainbow Crosswalks," *Huffington Post*, November 15. www.huffingtonpost.ca/entry/ chilliwack-rainbow-crosswalk-pride_ca_5dcf2ee3e4b0d2e79f8cd358.

Zanin, Andrea. 2017. "Your Country Needs You." In *We Still Demand!*, edited by Patrizia Gentile, Gary Kinsman, and L. Pauline Rankin, 185–202. Vancouver: UBC Press.

Suggestions for further study

- *Archives of Lesbian Oral Testimony (ALOT)*. A virtual archive compiled by E. Chenier. www.alotarchives.org.
- *Queering the Map*. A crowd-sourced map of LGBTQ+ stories. www.queeringthemap. com/.
- *United in Anger: A History of ACT UP*. A documentary film directed by Jim Hubbard. Produced by United In Anger, Inc. www.unitedinanger.com. 93 minutes.

4

PEOPLE WITH DISABILITIES

Enabling the city and planning for inclusion

Louise C. Johnson, Richard Tucker, and Valerie Watchorn

Introduction

The history of planning in Australia, as in other parts of the Anglosphere, is most usually linked to the creation of the profession, systematized bodies of knowledge and planning schools. Planning critiques from the 1960s around the exclusion of minorities, women and the environment from planning involved challenges to the planner as expert. This wave of critique also emanated from those with lived experience of disability, producing changes in building access and communications technologies as well as housing design, and the configuration of entire urban environments. However, integrating the history of defining, rendering services, and spatially regulating those with lived experiences of disability does complicate and broadens this planning history far beyond these recent responses.

A health impairment is a factual entity, but a disability is a social construct (Braddock and Parish 2001: 12). For those who were variously designated as not having the physical or mental capacity to live or work as others from the middle of the 19th century in Australia, the physically "disabled" and those with intellectual disabilities – designated "crippled," "deaf, dumb, and blind," "lunatics," and "idiots" – were planned for through an array of special spaces and institutions as well as regulatory regimes. Such interventions supplemented but did not supplant earlier realms of caring and self-sufficiency within family or street settings (Gleeson 2001a). From the mid-19th century, large, segregated institutions were planned and built, incarcerating those with physical and intellectual disabilities as well as mental illness together. Here the minimal care was transformed in the early 20th century by more interventionist eugenic therapies.[1]

From the late 20th century, this system was destabilized by a series of scandals around neglect but also new ideas of mental health and mechanisms to enhance physical capacity. Subsequently challenged by those with disability and their

DOI: 10.4324/9781003157588-5

advocates, and within the context of a global rights movement from the 1960s, spatial segregation and institutionalized models of care were largely dismantled. However, it is arguable that earlier institutions have been replaced by others, despite the many positive developments which have occurred. Disability, therefore, is still something that arises because access and inclusion is restricted and still marks individuals and groups for distinct regulatory and spatial consideration – through group homes as well as building modifications. It is these various interventions, dating from the earliest years of the colony up to the present day, which now need to be written into the history of Australian planning.

Late 1800s–1900: colonial planning for disability

It is unclear or unknown how Indigenous populations defined or dealt with those individuals who, in today's terms, had physical impairments or intellectual limitations. Some examples exist of ready accommodation of all into clans, though there are also examples where physical impairment – as result of violence, accident, birth, or old age – led to abandonment as the group moved on to the next camping site (Bonwick 1856). Linguists reported that signed communication was used within some language groups for specific purposes, like mourning periods or while hunting, suggesting the ready inclusion of those with hearing or sight impairments (Carty 2017).

But once colonization began, Indigenous peoples were progressively dispossessed of their country and confined to protectorates, reserves, or into institutions. Within the colonial city and in these fringe camps there was an increased likelihood of disease and poverty leading to homelessness and therefore life on the street for Indigenous peoples. Here they joined others who were poor and those who had physical or mental impairments that meant they were unable to work.

Australia had imported models, institutions, and ideas for virtually all aspects of life directly from England, though there was a rejection of the Poor Law model of government support via taxation, and care was rendered primarily within families or via charities (Gleeson 1995). Distinctions were made between "idiots" – persons with mental illness often framed as dangerous – and "invalids" – the only deserving group who were unable to contribute due to illness or age. Early on in the history of Sydney, "lunatics" and "idiots" were imprisoned on hulks, before the building of the first lunatic asylum at Castle Hill in 1811 (Carling-Jenkins and Sherry 2014).

Imported then were models by which those who were deemed physically or mentally different were to be accommodated. In 19th century Britain, deaf adults had begun to gather for social activities and religious worship from the 1820s, usually led by hearing people with evangelical or educational connections. These gathering places evolved into charitable organizations variously called "missions," "institutes," "benevolent societies," or "associations." Missions with close connections to the established Protestant churches were the model for early Australian deaf societies.

If not Indigenous, those with conditions such as blindness, deafness, or being unable to speak could become integrated into colonial society. Thus, there were deaf convicts as well as free settlers, one – John Carpenter – becoming a well-known artist and engraver in Sydney and another a successful pastoralist. In the early years of Sydney, Hobart, and Melbourne, those variously limited physically were not apparently segregated, though they may well have been consigned to the street if families were unable or unwilling to care for them (Gleeson 2001a).

As the 19th century progressed, those with physical limitations and mental impairments increasingly existed within a post-Enlightenment world view that had transformed the relationship between humans, nature, and God and elevated the role of science, the physician, educator, and caretaker. While there was official recognition that an intellectual disability was different from a mental illness, in reality the two were merged. As a consequence, there developed detailed definitions, diagnostics, treatments, and services for what was now constructed as a group on the wrong side of a new boundary between normalcy and aberrance (Braddock and Parish 2001). Whereas those with mental health issues could well end up in gaol, from the 1840s there was a series of Acts which were designed to define and manage "lunacy" in dedicated places. These were institutions that were large and in often expansive grounds – the idea being that contact with nature was therapeutic but also that separation from the city was vital – while the buildings themselves were formidable and closed. Asylums were significant in the urban landscapes of 19th century Australian cities, their location and operation planned for and regulated by the colonial state (via Boards and dedicated Acts).

If not incarcerated, those deemed physical "freaks of nature" – especially those with overly large or small stature, hair growth or physical features – were often displayed in an array of side shows and exhibitions across Britain, Europe, the United States, and Australia (Bogdan 1988; Broome and Jackamos 1998). As the Farmers Advocate noted in 1923: "there is always a crowd, ready to pay hard-earned money to see some freak of human nature" (quoted in Boumans 2019) – what Robert Bogdan (1988) describes as "the pornography of disability." Thus, in the early years of Australia's "settlement," the mentally disabled had been delimited, confined to their homes and the care of family members or on the edges of society, in "freak shows" and on the streets; but as the century progressed, they were increasingly incarcerated in dedicated institutions within cities as well as adjacent rural areas.

While people with intellectual impairments were legally and spatially separated from those with physical impairments, there was an issue within the medical and regulatory systems of the distinction between those with a mental illness (which was seen as temporary, treatable, and curable) and those with intellectual disabilities (seen as life long and incurable). While there was an awareness that such a distinction existed, as the 19th century progressed, it became blurred, such that "idiocy" and "insanity" were deemed comparable. In general, both groups were considered together under labels such as "idiots" and "lunatics" with their lives severely regulated through various "lunacy" statutes,[2] which defined these conditions and assumed their incarceration in special housing and hospitals.[3] Intellectual disability was therefore considered a form of insanity.

The different ways in which men and women were diagnosed, treated, and regarded within such institutions was further limited by gendered stereotypes. At least some patients also had physical disabilities, and these became part of the insanity diagnosis (Coleborne 2018). The women's conditions of "hysteria" or other supposed illnesses were often associated with sexual assault, sexual transgressions, violence, poor nourishment and neglect (Coleborne 2018). In Western Australia from 1829 till 1857 "lunatics" went to gaol or the colonial hospital where, if needed, they were then restrained. After that they went to the Fremantle Asylum, ultimately a space in which to lock people up who were socially deviant – the poor, sexually deviant, "Asiatics," or those who expressed unacceptable social behaviors, including being unable or unwilling to work. By 1896 the Fremantle Asylum had 190 inmates. Once placed in the asylum, few ever left (Cocks et al. 1996: 48). In contrast to the harsh treatment of women, the institutions sought to rehabilitate European men for employment, focusing their energies on who could recover and be released (Matthews 1984). The provision of "services" then, one of the core elements of any planning system, was highly gendered in the case of mental health facilities across the 19th and into the 20th centuries and delivered within tightly planned institutions.

Within the Australian city – born commercial but quickly becoming industrial – the separation of home from work – the primary spatial relationship of capitalism – was "a major mobility barrier for disabled people and undermined their capacity to get employment" (Gleeson 2001a: 225). Mechanical forms of production also assumed able-bodied workers to thereby render whole categories of the population – orphans, widows, the elderly, and those with physical impairments – unable to participate. Some institutions were specifically established in an effort to train in alternative skills those who could not readily enter the economy. Thus in 1860 in Sydney as well as in Melbourne, following the English model, institutes were established for the "deaf, dumb, and blind"; teaching and using manual communication to educate their charges (Carty 2017). Comparable institutions were established in South Australia in 1874 and Western Australia in 1896, before which deaf children from these colonies were sent to New South Wales and Victoria for their education. The Catholic Church also established schools for deaf children though it was not until the early 20th century – after sustained advocacy – that their education became compulsory.

As a consequence, those with intellectual but also physical impairments joined the poor and the sick in various spaces of exclusion – workhouses, hospitals, and asylums – run primarily by private charities bolstered by state subsidization. Those unable to do paid work received either "indoor relief" – residential care in various institutions – or "outdoor relief" which was provided directly to the home by volunteers from various charities. As well, of course, there was the care offered and provided by families. Those who did not have access to any of these systems or who accessed them sporadically, spent much of their time on the streets: as hawkers, beggars, and entertainers in "freak" shows.

In the early years of the Australian colonies, those with mental or physical disabilities were variously accommodated – within families and on the streets but also within prisons or in fringe camps and reserves if Indigenous. As the 19th century and urbanization progressed, the importance of the distinction between those who were able and those who were not able to engage in paid work grew, leading to systems to "manage" the deserving and undeserving poor with indoor and outdoor relief. If planning is in part about the delivery of human services to those in need, then it was charities who delivered some support to those who needed them but also gaols and asylums which physically removed those with intellectual and physical disabilities from the general population into large, well planned, and managed institutions. All of this occurred through state regulation well before the formal creation of planning. With the arrival of new models of "care" for those with mental impairment – be it a result of mental illness, "hysteria" or intellectual disability – there were major institutions constructed for their incarceration and "management" which, as the new century opened, involved far more intrusive and judgmental actions.

1900–1945: health, eugenics, and segregation

While schools did exist for those who were deaf, dumb, and blind, it took until the early 20th century for the education of these children to become compulsory (for the able bodied, secular, free, and compulsory education dated from the 1870s) and for organizations and services for deaf adults to emerge. Modeled on British missions, they were to provide religious counsel, regular sporting, educational and social activities. But "an implicit function of the missions was also to keep deaf people off the streets and out of undesirable places" (Carty 2017, 12).

By the turn of the 20th century people with an intellectual disability and mental health issues were at last recognized as different to one another, but both groups were still institutionalized together. With eugenic ideas ascendant, "disabled" Australians were routinely sterilized (Carling-Jenkins and Sherry 2014). In 1926 the Race Improvement Society was set up and in 1928 the Racial Hygiene Association. The Medical Journal of Australia openly recorded the medical fraternity support for sterilizing "mental defectives" in 1931 and it was not until 1960 that the Racial Hygiene Association changed its name to the Family Planning Association.

Meanwhile, the numbers of those needing assistance rose dramatically, primarily as a result of the two World Wars which saw many soldiers return with horrific physical injuries but also enduring mental challenges – collectively described as "shell shock." Because of the veneration of the "digger" as war hero and the embodiment of the Australian national character, veterans had to be taken more seriously than those with mental illness before. There was also a number of polio epidemics – the first in 1895 but then a series in 1936–1938, 1945–1947, 1951, and 1954 – which greatly increased the numbers of children who acquired physical impairments. Meanwhile, medical science was advancing, though it took until 1941 to establish the link between rubella and disability in Australia. As a consequence, there were some dedicated government services and limited financial support provided, though families were still expected to assume most care and existing institutions continued to incarcerate those with mental impairments (Carling-Jenkins and Sherry 2014).

1945–1980: deinstitutionalization and the disability rights movement

For people with disabilities, significant change in their perception and treatment arose from a collective consciousness of being excluded, along with a sense of empowerment, the embrace of the human and civil rights perspective along with international agendas, models, and organizations (Cooper 1999). Fundamental to this change was a shift in the international environment. The United Nations Declaration of Human Rights in 1948 initiated a worldwide rights movement. From World War II there emerged an array of organizations of those with physical impairments, their parents and carers who first gathered together to swap experiences and then demanded improvements in services and the environments in which their children lived and worked. There were very public exposés of the conditions within some of these institutions – including the 1966 Blatt and Kaplan's photographic essay on living conditions of those with intellectual disabilities in the United States and the 1962 novel by Ken Kersey which in 1975 was made into the widely seen film "One Flew over the Cuckoo's Nest" (Finnane 2009; Wiesel and Bigby 2015). All of this generated public outrage and calls for change. There was subsequently a series of international conventions that affirmed the rights of "the disabled": in 1971 Rights of Mentally Retarded Persons, in 1975 United Nations Declaration on the Rights of Disabled Persons, and in 1981 the International Year of Disabled Persons – which had as its aim "full participation and equality."

Institutions housing those with mental illness (and disability) also became increasingly unviable as a result of rising costs – wages, building, and grounds upkeep – crowding and the development of alternative and increasingly palatable ways of supporting people beyond their walls. New ideas and models of care were also being developed internationally along with increasingly strident criticisms of existing institutions. In Britain and Scandinavia, smaller scale residential units, early intervention programs and behavioral therapies all provided viable alternatives while social movements led by parents and people with lived experiences of disability demanded an end to incarceration within dedicated institutions. There developed a movement based on the civil rights discourse for deinstitutionalization and community-based care. However, the shift away from institutionalization was slow and haphazard as well as contested, with parents known to demand improvement of the system rather than its dismantling and governments committing to the process but not adequately funding it (Wiesel and Bigby 2015).

As Gilligren (1996) notes, despite the rights movement and more liberal approaches to treatment coming from overseas, public sector residential care right up to the 1960s was based on the assumption: "once a defective, always a defective" and occurred in dedicated, segregated and planned large-scale institutions, often in the middle of bustling cities. However, from then things began to more fundamentally change. In 1963, a reformist Labour government created the Handicapped Persons Assistance Act and even suggested a national disability insurance scheme, while the Henderson Poverty Inquiry specifically sought data on disability, identifying "handicap as the greatest single cause of poverty in Australia" (quoted in Carling-Jenkins and Sherry 2014, 54).

1980–2010: activism, Better Cities program, and legal reforms

If the first wave of deinstitutionalization occurred in the 1960s and 1970s, it involved mainly the downsizing of institutions and the piecemeal relocation of residents. But a new wave of activism, legislative change, and urban reform began in the 1980s. Thus, in 1983, Disabled People's International (Australia) was established and the Australian states started more systematically to deinstitutionalize and create community-based services while also moving away from the medical model of "disability" to a social model. In 1985, the Handicapped Person's Review called for anti-discrimination legislation and delimited State and Federal responsibilities, with changes to income support and funding directed to self-help groups, including home and community care programs. The 1986 Disability Services Act saw deinstitutionalization and "community care" with significant funding coming from the federal government.

In Victoria, while deinstitutionalization had been recommended since the 1970s,[4] only one institution had been closed by 1985 (St Nicholas for Intellectually Disabled children). In 1988, the state initiated a ten-year plan for the redevelopment of services for the "intellectually disabled," the closure of all institutions and the relocation of residents into community-based services, especially into group homes. Despite such statements, however, by 1989 two thirds of disability services were still delivered via institutions continuing to house over 2,600 people with intellectual disability (Wiesel and Bigby 2015). The Federal government became more actively involved in urban planning from 1991 to 1996 when the Better Cites agenda saw deinstitutionalization as a key element but also part of a national urban policy agenda. Funding within this program did allow the more rapid closing of institutions,[5] with their residents all moved into community-based housing (Wiesel and Bigby 2015).

Despite these actions across Victoria, there were still four state run institutions in 1999, including the very largest one: Kew Cottages in Melbourne. Its redevelopment highlighted the planning issues involved in a very large centrally located site. Here there was the Parent's Association who favored on-site cluster housing, disability rights groups advocating closure, and others, including land developers and the council, who saw an opportunity to build high density but also quality housing on the site. The final plan involved a planning compromise, with a new upmarket suburb developed on the site, with the sale of these dwellings funding the rehousing of the previous residents; cluster housing for the displaced in a small area of the new suburb, 20 group homes for 100 former residents and the remainder of the residents dispersed across the city in group homes (Wiesel and Bigby 2015). The location of these group and community homes, while permitted in all residential zones, often elicited spirited resident resistance, keeping the issue of housing those with disabilities at the forefront of the planning system, emphasizing their special status, their separateness and their collective forms of housing that necessarily marked them off from others.

By 2000, and following the 1992 Disability Discrimination Act, the old idea of a national disability insurance scheme was revisited. Long-term advocacy by carers, family members, and those with disabilities culminated in the ratification of the UN Convention on the Rights of People with Disability (UNRPD) in 2008. The Convention committed Australia to community inclusion, independent living, access to justice, freedom from abuse, health, and education. This was subsequently echoed in a range of disability Acts across the country as well as disability action plans at Federal, State, and Local government levels. Milner and Hayward (2017, 30) point to an "extraordinarily successful public campaign" by skillful leaders who healed rifts between disability advocates, carers, and service providers and responded systematically to the well-funded consultation processes that accompanied various enquiries into disability. The preparation of the National People with Disabilities and Carer Council's 2009 Shut Out Report was followed by a Productivity Commission Inquiry into Disability Care and Support (2011), which confirmed the many problems raised by activists – of underfunding, fragmentation, unequal, and ineffective care (Productivity Commission 2013). These reports and advocacy, which highlighted the marginalization, medicalization, and infantilization of those with a disability, were met by a bi-partisan call and actions for change.

2010–present: current plans, Universal Design, access, and inclusion

Out of these reforms and activism emerged the National Disability Insurance Scheme (NDIS, from 2013), which promises to revolutionize the care and support of those designated as "high needs" in Australia; by locating responsibility and funding at the highest Federal level of government, delegating service delivery to competing agencies and the power of choosing what is needed along with adequate resources to those most in need (Wiesel et al. 2017). The emphasis of funding has therefore shifted from dedicated institutions to the clients themselves via a person-centered model (Green and Mears 2014). The scale is massive – estimated at A\$22 billion in 2013 (David and West 2017: 331) with a huge bureaucracy – of 3000+ employees – all overseen by the Federal government. The NDIS is based on converting clients and those in need into consumers who have to negotiate their own service package, every year, on the basis of their need. Further, they have to deal with a now proliferating range of service providers who compete among themselves on the basis of quality but also cost, generating a pressure to casualize, anonymize, and cut the wages of those they employ (Coulter 2019; David and West 2017). The NDIS is therefore both a central government revolution in disability care but also consistent with many neo-liberal imperatives to individualize service access and commercialize service provision across space rather than at particular centralized locations.

The planning implications of the NDIS mesh with the shift away from the centralization of those with disabilities in institutions. People with different needs are therefore no longer physically concentrated or service provision centralized. Definitions of need remain codified and bureaucratized while formal housing remains potentially segregated, albeit small scale and self-managed, encountering the planning system in the location of group and community clusters. However, there are also significant changes in the context in which such planning is occurring, with local governments actively concerned to enable access and inclusion for all.

Design approaches that aim to be inclusive of the needs of the largest range of users possible are described by a range of terminologies in the literature, including "Design for all," "Barrier free design," "Universal Design," and "Inclusive design" (Deardorff and Birdsong, 2003; Persson, Åhman, Yngling, and Gulliksen, 2015). Universal Design has come to have the greatest currency in the context of Australian built environments; a development reflecting a broader, intersectional understanding of accessibility and inclusivity.

Although Universal Design originated to support participation of, and reduce discrimination against, people with disability, at its core Universal Design aims to enhance usability of built environments for all people, not just a minority group. Universal Design recognizes that the social participation of many people is reduced by environmental barriers and that people's interaction with environments is impacted not only by lifelong health conditions but also short-term conditions such as illness, injury, pregnancy, and age. By applying Universal Design to the early design of built environments, barriers can be avoided and people's participation maximized. To guide designers in the application of this concept, there are seven Principles of Universal Design (Connell et al. 1997): (1) Equitable use; (2) Flexibility in use; (3) Simple and intuitive use; (4) Perceptible information; (5) Tolerance for error; (6) Low physical effort; and (7) Size and space for approach.

As a means of achieving greater social inclusion and addressing the needs of increasingly diverse and aging populations, Universal Design is called for in a range of federal and state policies and, internationally in the United Nations Convention on the Rights of Persons with Disabilities (2007). However, the application of Universal Design is not mandated in Australian cities, and the design of Australian built environments continues to rely on a codified system of minimum accessibility standards and have built into them centuries of forms and modes of operation – such as car-oriented, sprawling suburbs – that ignore the needs of those with disability.

Conclusion

How does this history of disability intersect with the standard history of Australian planning? Clearly there are touch points, where early moves to plan institutions for those with intellectual disabilities (often merged with those considered mentally unwell) occurs at the same time as the early Australian cities were laid out or extended over the 19th century, as well as during the period of critique of 'who planners were planning for' from the 1960s which triggered the beginning of deinstitutionalization. So too there is a clear intersection point during the early 1990s when the Better Cities program of urban reform funded services and housing for those with disabilities and again with the National Disability Insurance Scheme as part of a neo-liberal reform agenda to privatize and individualize services. The Case Study included in this chapter also affirms the value of participatory planning along with a systems approach to the city.

But there are also significant points of divergence, primarily in the systematic moves by the colonies and States to physically separate, indeed to incarcerate those with different forms of "disability" and impairment; there to ensure their regimentation, service delivery and management as a sub-group of the population deemed lesser. It was against these planning regimes that the international moves toward human rights and the actions of parents and disability activists assume a vital role in altering the dominant regimes of physical management and service delivery. The results are moves toward Universal Design, access, and inclusion and the unfettered provision of services to those in need. The issue of system wide-understandings and interventions remain to be resolved by the planning and other social welfare systems.

Alternative planning history and theory timeline

Period	Planning history and theory canon	Alternative timeline: people with disabilities
Late 1800s–1900	Birth of planning	*Colonial planning for disability* Home-based care, "Indoor" and "Outdoor relief" for the poor, old, infirm, "freak shows" and survival on the street, distinctions between mental and physical impairment, segregation of the mentally ill/impaired
1900–1945	Formalization of planning	*Health, eugenics, and segregation* Institutionalization and socio-spatial marginalization of the "disabled" continues, eugenics and new levels of intervention with forced and voluntary sterilization
1945–1980	Growth/midlife crisis of planning	*Deinstitutionalization and the disability rights movement* Welfare capitalism, car-dominance produces personal mobility barriers, the disabled are still in institutions but are increasingly gaining a voice via associations
1980–2010	Maturation of planning	*Activism, Better Cities program and legal reforms* UN declaration of the Rights of Disabled Persons (1975), UN Charter of Rights, activism within the disability community and allies, deinstitutionalization part of the Better Cities urban agenda (1991–1996)
2010–present	New planning crisis	*Current plans, Universal Design, access, and inclusion* National Disability Insurance Scheme, community-based housing challenges, Universal Design Access and Inclusion on local government agendas, exposes of abuse and neglect in Royal Commissions

Case study: Accessible & Inclusive Geelong

Geelong is the second city of the state of Victoria, Australia (after Melbourne). In 2019, the Victorian Department of Health and Human Services commissioned a multi-disciplinary research group to determine the feasibility of making Geelong a "world-class" accessible and inclusive city. The aim was ambitious, but a regional city with an already-engaged community who for many years had lobbied for inclusion, visibility, and change, presented a unique context to realize this end. However, lack of progress over this time suggested the need for a more participatory and holistic process. There was a need to recognize the city as a complex system of interrelated parts, including not only aspects of urban planning, housing and building design but also community infrastructure and the processes and structures limiting employment and economic participation. Systems thinking was adopted as a participatory method 'for involving communities in the process of understanding and changing systems from the endogenous or feedback perspective of system dynamics' (Hovmand, 2014, p. 1). The method is rooted in the principles and approaches of participatory research; building on work addressing complex health issues, how community factors are interrelated, and what intervention can best address these. The process was guided by regular inclusion of the voices of those largely excluded historically – those with lived experiences of disability – and was underpinned by a view that promoting social inclusion for people with disability aligns with promoting social inclusion for all people due to the intersectionality of disability with other characteristics of diversity. Participants represented a wide spectrum of disability experiences. Two modes of primary data collection were used: workshops that utilized the Systems Thinking in Community Knowledge Exchange (STICKE) tool; and focus groups to refine the outcomes of the STICKE workshops to ensure they were user centered. Data from the workshops consisted of two primary forms: confirmed systems-maps; and priority actions. Three pervading themes were common across the systems-maps: Attitudes toward inclusion/access; Ability pride and inclusion as a value; and Valuing inclusion. Participants used the systems-maps to generate ideas for action to overcome obstacles to change. In total, 119 ideas across the workshops were suggested, with 34 of these prioritized according to their likely impact and feasibility. Identified ultimately were five Principles of Action and six Priority Actions. These included an affirmation of the importance of Universal and co-design, and the permeation of "inclusion" as a vital shaper of public policy, affordable housing, a Visitors Centre and employment. Twenty-eight interrelated actions were also recommended and grouped according to their alignment with each of the Priority Actions. These can be seen in full in the Accessible & Inclusive Geelong Feasibility Study linked to in the *Suggestions for further study*. In conclusion, the study brought together a plethora of research, strategic plans, diverse lived experiences and expertise to forge new collaborative relationships. Outcomes were enhanced by the transformative nature of the knowledge exchange and creation between participants. As a result, the changes suggested and embraced by decision makers and the community should be sustainable. Importantly, the systems approach underlined that the actions recommended in the collective plan cannot occur in isolation, because they can only overcome systemic lassitude by being implemented in combination to effect real and long-lasting change.

Notes

1 Emerging from the application of Charles Darwin's theory of evolution to human populations toward the end of the 19th century, Eugenicists believed that disability was physiological and inherited, to thereby sanction interventions in human reproduction to enhance desirable human characteristics while suppressing others. Thus, in early 20th century Australia child endowment and widow support was offered to white women but denied to Aboriginal mothers. Reaching its apotheosis in Nazis actions against particular population sub-groups – Jews, homosexuals, and the disabled – in Australia these views also underpinned actions to enforce the sterilization of those with disability while also limiting marriage and child rearing.
2 Via the 1860 Criminal Lunacy Act and 1867 in Victoria.
3 For example, the Kew Cottages in Melbourne (from 1865) and Claremont Hospital for the Insane in Western Australia (opened in 1907). Thus, the 1871 Lunacy Act clarified: "lunatic shall mean and include every person of unsound mind and every person being an idiot".
4 Via the Evans Report.
5 In Victoria Caloola (Sunbury), Mayday Hills (Beechworth) in 1992; Aradale (Ararat) in 1993, Janefield and Kingsbury (Melbourne) 1996, and Pleasant Creek (Stawell) 1998.

References

Bogdan R. (1988) *Freak Show: Presenting Human Oddities for Amusement and Profit*. Chicago: Chicago University Press.

Bonwick J. (1856) *William Buckley: The Wild Man and His Port Phillip Black Friends*. Melbourne: Geo Nichol.

Boumans E. (2019) "Entertaining Australia in the Interwar Years: Cultural Representations of Proportional Little Show People." *Limina* 25 (1): 29–50.

Braddock D.L. and Parish S.L. (2001) "History of Disability." In G.L. Albrecht, et al. (eds.) *Handbook of Disability Studies*. Thousand Oaks, CA: Sage Publications, pp. 11–68.

Broome R. with Jackamos, A. (1998) *Sideshow Alley*. St Leonards: Allen and Unwin.

Burgmann V. (1993) *Power and Protest: Movements for Change in Australian Society*. St Leonards: Allen and Unwin.

Burton M. (1996) "Introduction." In E. Cocks et al. (eds.) *Under Blue Skies: The social construction of intellectual disability in Western Australia*. Perth: Centre for Disability Research and Development.

Carling-Jenkins, R. and Sherry M. (2014) *Disability and Social Movements: Learning from Australian Experiences*. Farnham, UK: Ashgate.

Carty B. (2017) *Managing Their Own Affairs: The Australian Deaf Community in the 1920s and 1930s*. Washington: Gallaudet University Press.

Cocks E. et al. (eds.) (1996) *Under Blue Skies: The Social Construction of Intellectual Disability in Western Australia*. Perth: Centre for Disability Research and Development.

Coleborne C. (2018) "Disability and Madness in Colonial Asylum Records in Australia and New Zealand." In Rembis, M., Kudlick, C., and Nielson, K. (eds.) *Oxford Handbook of Disability History*. https://doi.org/10.1093/oxfordhb/978019234959.013.17.

Connell B. R., Jones, M., Mace, R., Mueller, J., Mullick, A., Ostroff, E., Sanford, J., Steinfeld, E., Story, M., and G, V. (1997). *The Principles of Universal Design – Version 2.0*. Retrieved from https://projects.ncsu.edu/design/cud/about_ud/udprinciplestext.htm.

Cooper M. (1999) "The Australian Disability Rights Movement Lives." *Disability and Society* 14 (2): 217–226.

Coulter E. (2019) "Move to NDIS Resulting in Mental Health Worker Job Insecurity, Tasmanian Peak Body Says", abc.net.au Just In, July 24.

David C. and West R. (2017) "NDIS Self-Management Approaches: Opportunities for Choice and Control or an Uber-Style Wild West?" *Australian Journal of Social Issues* 52: 331–346.

Deardorff C. J. and Birdsong C. (2003). "Universal Design: Clarifying a Common Vocabulary." *Housing and Society* 30(2): 119–138. https://doi.org/10.1080/08882746. 2003.11430488.

Finnane M. (2009) "Opening Up and Closing Down: Notes on the End of an Asylum." *Health and History* 11 (1): 9–24.

Foucault M. (2006) *History of Madness.* London: Routledge.

Freestone R. (2010) *Urban Nation: Australia's Planning Heritage.* Collingwood: CSIRO Publishing.

Gilligren C. (1996) "Once a Defective, Always a Defective: Public Sector Residential Care, 1900–1965." In E. Cocks et al. (eds.) *Under Blue Skies: The Social Construction of Intellectual Disability in Western Australia.* Perth: Centre for Disability Research and Development, 153ff.

Gleeson B. (1995) "A Space for Women: The Case of Charity in Colonial Melbourne." *Area* 27(3): 193–207.

Gleeson B. (2001a) "Domestic Space and Disability in 19th Century Melbourne, Australia." *Journal of Historical Geography* 27 (2): 223–240.

Gleeson B. (2001b). "Disability and the Open City." *Urban Studies* 38(2): 251–265.

Gleeson B. and Low N. (2000) *Australian Urban Planning: New Challenges, New Agendas.* Sydney: Allen and Unwin.

Green J. and Mears J. (2014) "The Implementation of the NDIS: Who Wins, Who Loses?" *Cosmopolitan Civil Societies Journal* 6 (2): 25.

Gurran N. (2011) *Australian Urban Land Use Planning: Principles, Systems and Practice.* Sydney: Sydney University Press.

Hall P. (2002) *Urban and Regional Planning.* New York: Routledge.

Hamnett S. and Freestone, R. (eds.) (2000) *The Australian Metropolis: A Planning History.* Crows Nest: Allen and Unwin.

Hovmand P. S. (2014). "Group Model Building and Community-based System Dynamics Process." In *Community Based System Dynamics.* Berlin: Springer, 17–30.

Jackson S., Porter, L., and Johnson, L.C. (2018) *Planning in Indigenous Australia: From Imperial Foundations to Postcolonial Futures.* New York: Routledge.

Kroll T., Barbour R., and Harris J. (2007). "Using Focus Groups in Disability Research." *Qualitative Health Research* 17(5): 690–698. Retrieved from https://journals.sagepub. com/doi/10.1177/1049732307301488?url_ver=Z39.88-2003&rfr_id=ori:rid:crossref. org&rfr_dat=cr_pub%3dpubmed.

Matthews J.J. (1984) *Good and Mad Women: The Historical Construction of Femininity in 20th Century Australia.* Sydney: George Allen and Unwin.

McCarty B. (2017) *Managing Their Own Affairs: The Australian Deaf Community in the 1920s and 1930s.* Washington: Gallaudet Press.

Milner P. and Hayward D. (2017) "Social Policy 'generosity' at a Time of Fiscal Austerity: The Strange Case of Australia's National Disability Insurance Scheme." *Critical Social Policy* 37 (1): 128–147.

Mumford L. (1991) *The City in History: Its Origins, Its Transformations and Its Prospects.* London: Penguin Books.

National Insurance Disability Scheme (NDIS) (2015) National Disability Insurance Agency, *Annual Report 2014–15*, https://www.ndis.gov.au/html/sites/default/files/documents/Annual-Report/201415-Annual-Report1.pdf.

Parham S. (1993) "How Has Women's Involvement in Urban Planning Changed Our Cities?" *Urban Futures* 3 (3): 46–50.

Persson H., Åhman H., Yngling A. A., and Gulliksen J. (2015). "Universal Design, Inclusive Design, Accessible Design, Design for All: Different Concepts—One Goal? On the Concept of Accessibility—Historical, Methodological and Philosophical Aspects." *Universal Access in the Information Society* 14(4): 505–526. https://doi.org/10.1007/s10209-014-0358-z.

Porter L. (2010) *Unlearning the Colonial Cultures of Planning*. Aldershot: Ashgate.

Productivity Commission (2013) *Disability, Care and Support*. Canberra: Productivity Commission.

Sandercock L. (1998) *Towards Cosmopolis: Planning for Multicultural Cities*. Chichester: John Wiley and Sons.

Sanders T. I. (2008). "Complex Systems Thinking and New Urbanism." In *New Urbanism and Beyond: Designing Cities for the Future*. New York: Rizzoli, 275–279.

Whitzman C. (2007) "The Loneliness of the Long Distance Runner: Long-term Feminist Planning Initiatives in London, Melbourne, Montreal, and Toronto." *Planning Theory and Practice* 9(2): 205–227.

Wiesel I. and Bigby C. (2015) "Movement on Shifting Sands: Deinstitutionalisation and People with Intellectual Disability." *Urban Policy and Research* 33 (2): 178–194.

Wiesel I., Whitzman C., Bigby C., and Gleeson, B. (2017) "How Will the NDIS Change Australian Cities?" *MSSI Issues Paper No. 9*. Melbourne Sustainable Society Institute, The University of Melbourne.

Suggestions for further study

- Centre for Universal Design Australia: https://universaldesignaustralia.net.au/.
- Australian Network for Universal Housing Design https://anuhd.org/.
- Findlay, C. (2021). *Growing up disabled in Australia*. Black Inc. Melbourne, Australia.
- Ted Talk: Elise Roy – When we design for disability we all benefit. www.ted.com/talks/elise_roy_when_we_design_for_disability_we_all_benefit.
- Accessible & Inclusive Geelong Feasibility Study Final Report and project video – https://deakinhomeresearchhub.com/aig/.

5

OLDER ADULTS

The detached ageism of the planning profession

John Lewis

Introduction

The age-friendly communities movement was introduced by the World Health Organization in the early 2000s and entered the vocabulary of planning practice as a belated recognition that cities globally are rapidly ageing (WHO, 2007). By now, most Canadian planners are acutely aware that the population of older adults in many parts of the country will represent one quarter to one fifth of the urban populations by mid-century (Menec et al., 2015; Statistics Canada, 2019). Moreover, planners have been reminded that, in the absence of a focused effort to address the "demographic timebomb" (Green, 2013), "senior surge" (Hodge, 2008), or "silver tsunami" (Warner and Zhang, 2019), provincial/state and local governments will be challenged to provide essential services and, in particular, will burden younger generations who will be required to "shell out" for older adults without the prospect of it ever being reciprocated (Karpf, 2015).

The kind of apocalyptic demography arguments that have become a common refrain in planning scholarship for over a decade highlight a troubling and widespread unease with the aged and ageing – i.e., ageism (Butler, 1969). On one level, ageism is reflected in the oppositional thinking that has crept into public policy discourse, which sees access to services and public goods as a zero-sum game wherein whatever is given to older adults will ultimately leave less for the rest of the population. At a deeper level, ageist thinking and discourse is ingrained within North American culture to the point where it is not unfair to say that here youth is the embodiment of all that is valued – i.e., beauty, vitality, and creativity. To be old, on the other hand, is to be generally regarded as conservative, petulant, and impaired, both cognitively and physically.

Ageist thinking and behavior holds true even for the middle aged or those at the margins of being old. Although no longer as commonplace as it once was,

DOI: 10.4324/9781003157588-6

the expression "OK Boomer" has become a catch-all dismissal directed by youth and young adults toward people that are generally regarded as out of touch with current thinking and trends. Many consider this phrase to be an outward sign of intergenerational warfare (Kingston, 2020). More deeply, such tensions are an expression of the fears felt by younger people burdened by job, housing, and income insecurity and their animosity toward an older but more socially and economically privileged generation.

Beyond the way that society speaks about older adults, there is a very real and palpable dimension to ageist thinking that has been laid bare by the coronavirus pandemic. The pandemic has revealed the extent to which older adults are vulnerable and have been devalued as active members of families and communities. The prevalence of inhumane care and mortality within commercial long-term care facilities during the pandemic's first wave reflects what Pope Francis described as a "covert euthanasia." In other words, "[i]n this consumerist, hedonistic and narcissistic society, we are accustomed to the idea that there are people that are disposable" (Bergoglio and Skorka, 2013: 98). Much of what has been experienced and learned about ageism in the early 21st century triggers a basic question, such as "at what point did North America stop valuing seniors as elders"? More to the point, it is appropriate to ask, "what role has planning played as a contributor to ageism" and "what can planning do to address ageism"?

These interesting and relevant questions will be the focus of this chapter, which explores changing conceptions of age and ageing over the last two centuries, largely in North America, and the reframing of ageing as a "problem." The chapter posits that, since the mid-19th century, Western societies have come to view ageing not as an inevitable aspect of being alive but as one of life's problems to be solved through individual effort aided by the power of science and professional expertise. Planning has played a largely peripheral role in addressing the ageing problem, although it could be argued that, as a profession, it has been a factor in the emergence of ageist thinking but could contribute a great deal to address it.

Late 1800s–1900: nascent ageism

It is tempting to assume that changing attitudes toward ageing and older people could be explained primarily by the major societal transformations of the 19th century – i.e., urbanization, industrialization and the advent of modernist thinking – but these would only be partial explanations. In North America at least, transformations in age relations largely preceded the growth of large industrial urban centers. The explanation could be attributed initially to demographic changes as North America's population began to grow older and continued to do so well into the 20th century. Prior to 1810, less than two percent of North Americans could expect to live past their 60s, which grew to more than thirty percent of the population by the 1940s (Fischer, 1978). In so far as old age brought experience, and that form of lived experience and accumulated wisdom, was something quite rare in pre-industrial communities, it often conferred a measure of authority. The "scarcity value" of old age meant, on the one hand, that older adults were a rarity but it also meant that they would not be dependent on their families and the broader community (Karpf, 2015). For much of human history, old age had been an experience that only a limited few would encounter. However, in the span of two centuries, ageing has become more common and with it a growing sense of contempt that was reflected in the early 19th century and echoes into the early 21st century. That contempt and nascent form of ageism was and continues to be rooted in fear of deterioration and the attendant loss of vigor and beauty conferred by youth, as well as anxiety about access to resources and shouldering the burden created by older people (Cole, 1992).

By the time the Industrial Revolution came into its own in the mid-19th century, the explosive growth of urban populations took place in a laissez-faire era when the blind forces of chance and profit determined the form of cities and the conditions of urban life. While the physical landscape of the urban working class was changing for the worse, the economic and social conditions of older workers in particular were being transformed. Most noteworthy was the change in the relationship between age and employment and the advent of an institutionalized form of ageism through retirement. The growth of factory production unquestionably favored younger and more agile workers who could keep pace with the rapid pace of the assembly line. Dr. William Osler, widely regarded by medical historians as the founder of modern medicine, reflected the ageism (and sexism) of his time by proclaiming that men working beyond the age of 60 had reached the limits of their productive lives and that there would be "… incalculable benefit … in commercial, political and professional life, if as a matter of course, men stopped work at this age" (Osler quoted in McDonald, 2013: 8).

As an institution, retirement took root in industrialized economies in the late 19th century and became well-established by the end of World War II as a means of rationalizing industrial production by transitioning older and less productive workers out of the labor force. Sociologist Alan Walker has argued that, as a social construct, "… retirement is both the leading form of age discrimination and the driving force behind the development of ageism in modern societies" (1990: 59). In this sense, retirement has come to define clearly the boundary of old age as the final segment of an individual's life course, distinct from the formative stage of education and youth and the productive phase of employment, childbearing, and adulthood.

The pensions that retired industrial workers received were never intended to sustain them well or for the long term. The meager annuities that were paid through early pension schemes meant that retired workers were often consumed by fear − i.e., fear of poverty, illness, and fear of having to enter an almshouse or, at worst, a workhouse (Laws, 1993; Picard, 2021). Between the first state pensions in the 1850s to the early 1950s, the fear and anxiety of retired workers in North America was driven by recurrent economic depressions that wiped out whatever savings they had accumulated during their working years, and for which even the most generous pensions at the time could not provide adequate financial support.

1900–1945: positivist social reform

For the most part, scholars attribute the antecedents of ageism to the demographic changes and transformed nature of working relationships that emerged during the 19th century. However, contemporary theorists would also argue that the emergence of social reform movements that were rooted in the "philosophy of the 19th century" and intended to provide solutions to the problems of the industrial era reflected ageist thinking and, in some measure, contributed to it (Wilson, 2016). Any vague notions of town, country, or city planning scarcely cropped up throughout much of the 19th century, but it is clear that planning's origins can be clearly traced to a number of social reformers such as Robert Owen, Ebenezer Howard, Charles Booth, and Patrick Geddes influenced by Positivist sociology and modernity (Allmendinger, 2002; Wilson, 2018).

Yet the focus of the utopian visions articulated by the urban social reformers of the early 19th century to the planners of the early 20th was very much directed toward the living conditions of young working families (Beauregard, 1989). For the most part, family structures underwent major changes during that time to the extent that solitary residence increased among older retired male workers (Fischer, 1978). British and North American households were typically comprised of extended families or, where different generations did not live within the same residence, they were commonly very close to one another, often on adjacent plots of land. That pattern gradually changed to the point where, by the late 19th century, it became increasingly common for older adults to live alone. Whether the early planning profession's focus on young working families represented an inadvertent neglect or was simply a reflection of the wider Victorian mindset that old age represents dependence and decline, the needs of people who were no longer the focus of domestic family or industrial economic life were largely peripheral to the interests of the early planning profession and were rarely central to the utopian visions of modernist planning reformers.

That mantle would be taken up by another profession influenced by Positivist philosophy that would conceive of old age as a problem to be solved by the intervention of society through medical science (Tornstam, 1992). In the late 19th century, the founders of geriatrics and gerontology set about studying old age largely to discover the biologically determined or "essential" laws of senescence in order to ease the discomfort and pain of growing old (Weiss et al., 2016). A new medical focus on the pathology of ageing, growing populations of older retired workers in almshouses and middle-class fears about their own fate later in life likely served to accentuate ageing as the negative pole of the human life course. Although some scholars would contend that the reduction of the ageing experience to a medical problem has merely redefined older adults as patients in need of professional expertise and intervention (Cole, 1992), others would contend that the practice of gerontology as it grew in the early to mid-20th century would eventually come to challenge the myths reflected in popular sentiments about ageing (Butler, 1969; Thornton, 2002). It could also be suggested that the emergence of environmental gerontology as a sub-discipline in the 1950s sharpened attention toward, and criticism of, planning practice and its relevance for the ageing experience.

1945–1965: benign ageism

The planning profession's relevance waned during the Great Depression and World War II as the threats of material, political, and national insecurity rendered superfluous questions of efficient urban growth and social well-being. Planning's relevance would dramatically shift during the immediate postwar era of the late 1940s and 1950s bringing forth a suite of challenges that would, on the one hand, revitalize the planning profession but, on the other, lay the foundations for contemporary critiques of planning practice by environmental gerontology and age-friendly planning scholars. Particularly in North America, postwar reconstruction spurred unprecedented economic growth that produced rising incomes and unleashed a demand that had been constrained during long years of war and depression for housing and other, newer trappings of urban living – i.e., the private automobile. However, perhaps the most impactful phenomenon of the postwar period was the dramatic population growth or "baby boom" throughout much of North America. This propelled demand for housing, schools, playgrounds, and social services and, in turn, accelerated urban growth.

By the latter half of the 20th century, the dominant residential development pattern throughout North America would come to be rooted low-density, automobile-centered suburbs segregated from other land-uses. It was in this context that the environmental gerontology perspective began to take shape in the 1950s and 1960s (Carp, 1966; Kleemeier, 1959). The impacts of environmental gerontology in its early years were two-fold. First, it established a theoretical foundation that addressed the spatial context of ageing and argued that human development over the life course from childhood to old age is driven by an ongoing exchange or transaction between individuals and their social and physical environment (Hodge, 2008). Environmental gerontology's focus during the 1950s and 1960s was centered on the day-to-day or home context of ageing individuals. This was an early recognition that most home environments were not designed to address the changing physical and mental capacities of the occupants as they age. The ecological theory of ageing posits that old age is a critical phase in the life course of an individual and the capacity to live well and independently in later life is a function of "person-environment fit" or the dynamic interaction between the physical and mental "competence" of an individual and the "press" of his or her environment (Lawton and Nahemow, 1973). Competence is comprised of the capacities of an individual and consists of such processes as physical health, sensory and perceptual capacities, motor skills, cognitive abilities, and mental health (Lawton, 1982). Environmental press encompasses the characteristics of the physical and social environments and the challenges that those characteristics place on an individual. Difficulties arise as people in varying stages of the ageing process experience changing health, physical, and cognitive functioning such that their competence no longer matches the demands of their environment.

The second impact of the gerontology profession broadly was to contribute to the expression of a form of ageism, which laid the foundations for an industry focused on ageing and predicated on the assumption that there are generalizable limits to how much an ageing person and their environment can be adapted to overcome environmental presses. Lecturing to a gerontology class in 1943, George Lawton argued for what has come to be referred to as "benign ageism" to the extent that "If they [the elderly] cannot be given real lives, they must be given proxy ones" (George Lawton quoted in Cole, 1984: 364). The benign or compassionate ageism that found expression in the 1950s defined older adults as a singular group of people that are frail, dependent, and poor but are above all "deserving" (Binstock, 1983). This kind of stereotyping, unlike sexism or racism, was not entirely detrimental to the well-being of older people. For instance, in 1965 the Canadian government established the Canada Pension Plan, thereby assuming responsibility for providing a stable income to retired workers and effectively detaching the concept of retirement from poverty. Publicly managed pension programs became "deferred wages" to which people were entitled due to their contributions during their working lives, and which further served to delay if not prevent the withdrawal of older adults from economic activity. An entirely new industry took root at this time to tap the spending power of this new market by combining housing and medical care to provide a conception of sustained and active living to older adults through organized social encounters and medical services using novel terms such as "adult home," "retirement community," "assisted living," and "life care community" (Drucker, 2015).

Both of the newly formed professions of urban planning and gerontology were firmly rooted in 19th century assumptions concerning humanity's capacity to understand the world and, more importantly, grasp the problems of industrial change and leverage the power of scientific rationality to effect positive change. For the field of gerontology, the problem of chronological age was framed as a biologically determined and, therefore, value-free realm of inquiry. The benign ageism of social policy was driven by the assumption that older people had lost their capacity to contribute productively to the workforce and, as clients or patients, the aged ought to be set aside from the rest of society to receive the professional expertise and care that they deserve (Estes, 2001). The medicalization of ageing and the prevailing mindset of the 19th and early 20th centuries that age represents everything that bourgeois culture at the time hoped to avoid – i.e., dependence and disease – may account in part for a longstanding detached ageism of the planning profession. In effect, the needs and well-being of the aged were seen, either implicitly or explicitly, as peripheral to the expertise of the planning professional as the rational purveyor of built form to address the living conditions of young, productive working-class families. In the latter half of the 20th century, the problems that would define the professions of planning and gerontology would shift along with the waning modernist consensus that governed perceptions around the authority of professional expertise.

1965–1980: postmodern transformations

The late 20th century was dominated by challenges from "post" philosophies and critiques – i.e., postpositivist, poststructuralist, and postmodern. From the postmodernist perspective, since knowledge is inherently situational and unreliable, planners cannot claim to capture and define reality much less define solutions to the problems of the world. The postmodern era had a profound impact on the planning profession's focus and the role of the practitioner. Beginning in the 1960s, the postmodern era has recast the planner as a "fallible advisor" (Allmendinger, 2002) who functions in a world of complexity, of wicked problems where there are no clear solutions but only processes through which alternatives can be communicated and evaluated. The new focus of planning theory and practice on process and communicative rationality still clings to a modernist sensibility, albeit one in which the practitioner imposes their expertise on the process and communication of plans rather than as the visionary behind city form (Innes, 1995; Fainstein, 2014).

Postmodernism's influence on the scholarship and practice of gerontology began as well in the 1960s through challenges to the "essentialist" myths that were prevalent in the field's formative years of the early 20th century. In effect, contemporary attitudes toward ageing are rooted in the gerontology profession's search for underlying or essential biological causes that determine how all of us experience ageing (Weiss et al., 2016). Essentialist arguments have been used in studies of age, race, and gender to account for observed differences between men and women, for instance, in terms of biologically determined, universal causes. However, gender theorists have argued that, unlike sex, differences between what is considered "masculine" or "feminine" in any given society are not essential but socially constructed. In the same way that gender refers to a bundle of behaviors and practices that transform distinctions of sex into socially defined categories, societies have similarly transformed chronological age into social and cultural categories. For instance, the notion of "generations" was introduced by Karl Mannheim in 1928 and has come to define how society categorizes people based on chronological age and perceived capacities (Timonen and Conlon, 2015). How societies have defined the distinction between young and old has shifted throughout history and differs between cultures (Cole, 1992). Postmodern critiques of age-based essentialism posit that it is true that all people will encounter changes as their bodies age chronologically up to the point of death. However, the biological "fact" of ageing and the presumed limitations experienced as a result obscure the highly variable nature in which different individuals encounter the ageing process.

The latter half of the 20th century witnessed several movements which sought to challenge the racial and gender-based prejudices that had come to define how entire communities of people were described, how they were able to earn a livelihood and where they ought to live. The age-activist Maggie Kuhn founded the Gray Panthers movement in the 1970s to challenge the essentialist underpinnings of "disengagement theory" – i.e., that older adults are happier when they willingly withdraw from society and established social roles as their abilities deteriorate over time (Cumming and Henry, 1961). Kuhn openly confronted the widespread conception of older people as physically, mentally and, in particular, sexually incapacitated, and the segregation of older adults from mainstream society into "glorified playpens" (Woodspring, 2016).

1980–2000: the Boomers come of age

The 1980s and 1990s marked a time when the Baby Boom generation was emerging from their youth and into the realization that they could not count on the unprecedented prosperity that had defined their parents' adult and older adult years. The specter of an ageing society was admittedly overshadowed by fears of nuclear holocaust, social conflict, and environmental crises. However, the large numbers represented by the people born between 1945 and 1964 raised for the first time in popular media and policy discourse the possibility of intergenerational warfare through the "greying" of public budgets and the displacement of costs onto future generations (Cole, 1992). Much of this resonates into the present through the manifestly alarmist and ageist refrains about the "silver tsunami."

Alongside the resurgent ageism rooted in fears of scarcity, the 1980s and 1990s marked the emergence in post-modern gerontology of scholarship that focused on the spatiality of ageing and, in particular, on the built environment as "sites of struggle" for older adults (Harvey, 1990). Echoing Maggie Kuhn's challenges and calls to liberate older adults from age-based segregation, this early literature shed light on the paradoxical thinking prevalent in the decades following World War II. In effect, while it would be illegal and considered abhorrent to build segregated housing for people of a particular cultural or ethnic origin, people throughout the Anglosphere have gradually come to accept the idea of age segregated housing as something that is both socially acceptable and necessary. It is acceptable because, as consumers, the assumption has and continues to be that older adults generally wish to separate themselves from younger people and take on a slower pace of life. Age-segregated housing and elder care is admittedly a mode of living that has benefitted and is preferred by many older adults. However, to the same extent that it is inappropriate to assume that all people will experience ageing in generally the same way, it is unreasonable to suggest that all or most older people, given a more diverse spectrum of living opportunities, will invariably choose age segregated living. In addition to preference, it is argued that age-segregated housing is necessary to provide amenities and living spaces to people who require access to medical care services that can attenuate the problems of old age. The struggle according to postmodern gerontology is that age-based housing segregation is not only rooted in benign ageism but is itself a contributor to ageism. In other words, age-based stereotypes and prejudices are likely to exist and become commonplace in society when the prevalent response to ageing is to isolate older adults and limit opportunities to remain visible, socially interact with the broader community and maintain or establish social relationships. As Karpf (2015: 90) argues, "… young people's lack of contact with old people not only encourages them to believe that they'll never grow old, but also to treat people as though they'd never been young."

The years leading up to the new millennium saw a growing if perhaps belated acknowledgement that the sprawling, mono-use suburban neighborhoods planned and built to accommodate young Baby Boomer families can place their ageing occupants at risk when they can no longer drive to access services, amenities, or social networks, or when large dwellings become financially and physically difficult to maintain (Bookman, 2008). Influenced in part by neighborhood satisfaction research (Parkes et al., 2002), environmental gerontologists acknowledged that while the home environment may be adapted to accommodate the occupant's changing needs, the person-environment fit perspective should be broadened beyond domestic settings to recognize the wider community setting as a fundamental facet of maintaining older people's "… continuing participation in … social, economic, cultural and civic affairs …" (WHO, 2002: 12).

This perspective was influential in the development of ideas associated with lifetime neighborhoods (DCLG, 2008) and livable communities (Kihl et al., 2005). However, the flagship standard remains a program initiated by the World Health Organization in 2006 that captured the characteristics of an age-friendly city according to eight "domains": the built environment, transportation, housing, social participation, respect and social inclusion, civic participation and employment, communication, and community support, and health services. Defining an age-friendly community as one that "encourages active ageing by optimizing opportunities for health, participation and security in order to enhance quality of life as people age" (Plouffe and Kalache, 2010), the age-friendly cities movement has become the standard framework for a growing international network of communities to plan for an ageing Baby Boomer population.

2010–present: age-friendly community "planning"

The "AFC movement" has not been without its struggles. One issue has centered on branding or the relevance of age-friendly planning and challenging the conception that ageing is something that is endemic to the old – i.e., pensioners or people who reach the arbitrary benchmark of 65 years. In essence, ageing is an experience that every living person encounters from the moment of their conception to their ultimate passing. In ways that are broadly similar but fundamentally unique to each individual, our bodies and minds change throughout the life course, and urban and rural communities can be planned either to accommodate or exclude people from "… opportunities for health, participation and security …" at all stages of life, not just in the context of old age (Plouffe and Kalache, 2010). Part of the re-branding of age-friendly planning has been to suggest that by planning for the most vulnerable members of society, people of all stages of life and abilities will be the ultimate beneficiaries (Greenfield et al., 2015).

There is tremendous power and potential embodied in the AFC movement for planning. This is perhaps the most comprehensive framework describing multiple dimensions of built and social environments within urban communities that affect all people's quality of life across the lifespan. However, compared to gerontology and other health professionals, planning researchers and practitioners are conspicuously underrepresented at policy tables to plan and conceive what an age-friendly community (and city) could mean in terms of built form and social infrastructure. In large part, this may be due to the still prevalent conception both outside and within the planning profession that research or policy questions associated with human ageing are fundamentally a matter for health professionals. It may also reflect the persistence of postmodern thinking within the profession that has defined the planner's role as a communicator and advisor at the expense of a visionary who is capable of expressing the forms that communities may take to realize the content of age-friendly plans. The planning profession has an opportunity through the age-friendly framework developed by the World Health Organization to be entrepreneurial and visionary, to express multiple visions of what a better city comprises for people of all ages. More importantly, and perhaps in line with the logic of communicative planning, the planning profession has the opportunity to stimulate broader conversations about what it means to "age in community" and reverse the age-segregationist assumptions of the past decades (Thomas and Blanchard, 2009).

Conclusion

Returning to the questions that were posed at the outset of this chapter, there is no singular or straightforward explanation for the sources or causes of ageism. In broad terms, the emergence and prevalence of ageist thinking and behavior can be linked historically to demographic changes that rendered ageing more commonplace and less venerable; transformations in economic and working relationships that privileged younger, more agile bodies; as well as persistent essentialist tendencies to simplify the ageing experience and conceptualize it through a biological/medical lens that emphasizes dependence, increasing ill-health, disability, and decline (Landorf et al., 2008).

With the framing of ageing as a problem by gerontology in the late 19th to early 20th centuries, a new set of age relations emerged which embraced stereotypes of older adults as frail and dependent but deserving of support through government programs and the commercialization of housing and support services for the aged. The compassionate or benign ageism now prevalent in the ageing industry contrasts with the detached ageism of the planning profession. Detached ageism is essentially rooted in neglect and is reflected in the planning profession which has rarely viewed the challenges of ageing in industrial or post-industrial cities as fundamental to the bold new visions of urban life of early modernists, the postwar suburban spaces for burgeoning working families, or to the post-modern professional's focus on process and communication as the principal objectives of planning. Planning's detachment from the place of older adults in community life has affected or at least contributed to many of the problems that environmental gerontology and age-friendly scholars recurrently underscore as endemic to urban life for the elderly – i.e., suburbs designed for families and commuters on the one hand, and land-use policies that enable the segregation of older adults from the wider community on the other.

Planning's detached ageism is not deliberate to the extent that it is unlikely that the profession's practitioners and scholars have set about intentionally to remove older adults from the social fabric of communities. However, there is ample opportunity and need for planners to take up ageing as a relevant and important focus of professional and scholarly work. On one level, this would mean changing planning's role from advising other professionals how to formulate and communicate plans to envisioning what an age-friendly community means and could comprise for people from the beginning to the final years of life. If this sounds too much like returning to the top-down, utopian visions of planning's modernist roots then, on another level, planners could express more modest but nonetheless visionary models of living that provide more choice for older adults and fill the wide dichotomy that exists between ageing-in-place and age-segregated living. What "ageing in community" means, for instance, in terms of social policy and built form is surprisingly underexplored in planning scholarship. As a concept, ageing in community is rooted in providing inclusive and socially vibrant communities that provide access to resources, foster opportunities for multiple generations to share cultural and social encounters, and generally have a full life regardless of age or other circumstances (Thomas and Blanchard, 2009). There are limited examples of what this concept could entail from international sources but, in both theoretical and concrete terms, ageing in community remains largely underexplored as a concept and unrealized as a vision. Crafting multiple visions for ageing in community could provide a starting point for building conversations around how to produce Better Cities and towns for people across the life spectrum.

Ageing will likely remain a lightning rod for anxieties rooted in personal fears about growing old and general societal pessimism about the future. The practice and scholarship of planning will not play a leading or even a significant role in transforming how contemporary society views the aged. However, if the essential purpose of planning is to create the "just city" (Fainstein, 2005), then planning ought to do more than merely inform other professions how to plan for the aged. Under the pretense of doing what is ostensibly in the best interests of older adults, society has cleansed itself of the aged by creating "golden ghettos" (Mangum, 1979) that remove them from workplaces, learning institutions, communities and home settings. As Karpf argues, "It might soon be possible to go through life without meeting an old person until you become one. No wonder the prospect of ageing is terrifying" (2015, 89). Planning has the capacity as well as the opportunity to reverse the segregationist tendencies of modern society toward the aged and express the various forms that a just city for the aged could comprise, where older adults are fully integrated and visible contributors to the social fabric of a community.

Alternative planning history and theory timeline

Period	Planning history and theory canon	Alternative timeline: older adults
Late 1800s–1900	Birth of planning	Nascent ageism Social reformers, emerging ageism, demographic change, industrialization, urbanization, retirement
1900–1945	Formalization of planning	Positivist social reform Industrial working families, utopian master plans, ageing as problem, gerontology
1945–1965	Growth of planning	Benign ageism Neighborhood planning, baby boom, middle class families, suburbanization, compassionate ageism
1965–1980	Midlife crisis of planning	Postmodern transformations Complexity, essentialist challenges, disengagement, planner as rational advisor
1980–2000	Maturation of planning	The Boomers come of age Ageing boomers and suburbs, person-environment fit, neighborhood satisfaction
2010–present	New planning crisis	Age-friendly community "planning" Resurgent ageism, age-friendly communities, "silver tsunami," age segregation, pandemic

Case study: the gold standard in elder care

André Picard's recent book *Neglected No More* (2021), provides a timely polemic documenting the deplorable state of Canada's for-profit long-term care industry during the COVID-19 pandemic and posits that countries throughout the world have generally failed to care for our elders through a system largely based on mass institutionalization. During an interview about his book with the Canadian Broadcasting Corporation, when asked if there is any place in the world he could point to as "the model that we can apply here," without hesitation Picard states that Denmark embodies the "gold standard" (CBC, 2021). He suggests that the Danish system of elder care is fundamentally based on a national cultural mindset that values older adults as vital members of their community and where, from a policy perspective, segregation from the larger community is considered a last resort. For the most part, the Danish model is based on maintaining older adult independence and keeping elders within the community through substantial investments in home care, the delivery of "re-ablement" programs that teach older adults how to adapt daily activities to changing physical and sensory abilities and, where necessary, small-scale older adult residences are situated within residential neighborhoods and designed to be indistinguishable from adjacent residences. There is much that planners in North America can learn from the Danish approach to ageing in community. However, as Picard also contends, it will require policy makers to break from the established commercial/industrial model of elder care and that planners move beyond being the masters of writing reports and providing recommendations (i.e., advising and communicating). It will require planners to use their capacities to envision what ageing in community embodies, advocate for and implement that change.

References

Allmendinger, P. (2002). Towards a post-positivist typology of planning theory. *Planning Theory*, 1(1), 77–99.

Beauregard, R. A. (1989). Between modernity and postmodernity: The ambiguous position of US planning. *Environment and Planning D: Society and Space*, 7(4), 381–395.

Bergoglio, J. M. and Skorka, A. (2013). *On heaven and earth: Pope Francis on faith, family, and the church in the twenty-first century*. New York, NY: Image Publishing.

Binstock, R. (1983). The aged as scapegoat. *The Gerontologist*, 23, 136–143.

Bookman, A. (2008). Innovative models of ageing in place: Transforming our communities for an ageing population. *Community, Work & Family*, 11(4), 419–438.

Butler, R. N. (1969). Age-ism: Another form of bigotry. *The Gerontologist*, 9(4 Part 1), 243–246.

Carp, F. (1966). *A future for the aged: The residents of Victoria Plaza*. Austin, TX: University of Texas Press.

CBC (2021). André Picard on treating seniors with dignity after COVID-19 exposed a crisis. The Canadian Broadcasting Corporation. https://www.cbc.ca/player/play/1868103747511.

Cole, T. R. (1984). The prophecy of senescence: G. Stanley Hall and the reconstruction of old age in America. *The Gerontologist*, 24(4), 360–366.

Cole, T. R. (1992). *The Journey of Life: A cultural history of ageing in America*. Cambridge: Cambridge University Press.

Cumming, E., and Henry, W. E. (1961). *Growing old, the process of disengagement*. New York, NY: Basic books.

Department of Communities and Local Government. (2008). *Lifetime homes, lifetime neighbourhoods: A national strategy for housing in an ageing society*. London: DCLG.

Drucker, S. J. (2015). The zoning in and the zoning out of the elderly: Emerging community and communication patterns. *Proceedings of the New York State Communication Association*, 2014(1), 1.

Estes, C. L. (2001). *Social policy and ageing: A critical perspective*. Thousand Oaks, CA: Sage.

Fainstein, S. (2005). Planning theory and the city. *Journal of Planning Education and Research*, 25(2), 121–130.

Fainstein, S. S. (2014). The just city. *International Journal of Urban Sciences*, 18(1), 1–18.

Fischer, D. H. (1978). *Growing old in America: The Bland-Lee lectures delivered at Clark University*. Oxford: Oxford University Press.

Green, G. (2013). Age-friendly cities of Europe. *Journal of Urban Health*, 90(1), 116–128.

Greenfield, E. A., Oberlink, M., Scharlach, A. E., Neal, M. B., and Stafford, P. B. (2015). Age-friendly community initiatives: Conceptual issues and key questions. *The Gerontologist*, 55(2), 191–198.

Harvey, D. (1990). *The condition of postmodernity* (Vol. 14). Oxford: Blackwell.

Hodge, G. (2008). *The geography of ageing: Preparing communities for the surge in seniors.* Montréal: McGill-Queen's University Press.

Innes, J. (1995). Planning theory's emerging paradigm: Communicative action and interactive practice. *Journal of Planning Education and Research*, 14 (3): 183–189.

Karpf, A. (2015). *How to age.* New York, NY: Macmillan.

Kihl, M., Breenan, D., Gabhawala, N., List, J., and Mittal, P. (2005). *Livable communities: An evaluation guide.* Washington DC: AARP Public Policy Institute.

Kingston, A. (2020). "Why intergenerational warfare is a mug's game." *Macleans*, January 8, 2020. https://www.macleans.ca/opinion/why-intergenerational-warfare-is-a-mugs-game/.

Kleemeier, R. (1959). Behavior and the organization of the bodily and external environment, In J. Birren (Ed.), *Handbook of ageing and the individual* (pp. 400–451). Chicago: University of Chicago Press.

Landorf, C., Brewer, G., and Sheppard, L. (2008). The urban environment and sustainable ageing: Critical issues and assessment indicators. *Local Environment*, 13(6), 497–514.

Laws, G. (1993). "The land of old age": Society's changing attitudes toward urban built environments for elderly people. *Annals of the Association of American Geographers*, 83(4), 672–693.

Lawton, M.-P. and Nahemow, L. (1973). Ecology of the ageing process, In, C. Eisdorfer, M.-P. Lawton (Eds.) *Psychology of adult development and ageing* (pp. 619–624). Washington, DC: American Psychological Association.

Lawton, M.-P. (1982). Competence, environmental press, and the adaptation of older people, In, M.-P. Lawton, P. Windley and T. Byerts (Eds.) *Ageing and the environment* (pp. 33–59). New York, NY: Springer.

Mangum, W.P. (1979) Retirement villages: Past, present and future issues. In P.A. Wagner and J. M. McRae (Eds.), *Back to basics: Food and shelter for the elderly.* Gainesville: Center for Gerontological Programs and Studies, University of Florida.

McDonald, L. (2013). The evolution of retirement as systematic ageism. In, Brownell, P. and Kelly, J. J. (Eds.). (2012). *Ageism and mistreatment of older workers: Current reality, future solutions.* (pp. 69–90). Dordrecht: Springer Science & Business Media.

Menec, V., Hutton, L., Newall, N., Nowicki, S., Spina, J., and Veselyuk, D. (2015). 'How 'age-friendly' are rural communities and what community characteristics are related to age-friendliness? The case of rural Manitoba, Canada', *Ageing & Society*, 35(1), 203–223.

Parkes, A., Kearns, A., and Atkinson R. (2002). What makes people dissatisfied with their neighbourhoods. *Urban Studies*. 39(13), 2413–2438.

Picard, A. (2021). *Neglected no more: The urgent need to improve the lives of Canada's elders in the wake of a pandemic.* Toronto: Random House Canada.

Plouffe, L. and Kalache, A. (2010). Towards global age-friendly cities: Determining urban features that promote active ageing. *Journal of Urban Health*, 87(5), 733–739.

Statistics Canada. (2019). 'Canada's population estimates: Age and sex, July 1, 2018', *The Daily*, Friday January 25th, 2019. Available at: https://www150.statcan.gc.ca/n1/daily-quotidien/190125/dq190125a-eng.htm.

Thomas, W. and Blanchard, J. (2009). Moving beyond place: Ageing in community. *Generations*, 33(2), 12–17.

Thornton, J. E. (2002). Myths of ageing or ageist stereotypes. *Educational Gerontology*, 28(4), 301–312.

Timonen, V. and Conlon, C. (2015). Beyond Mannheim: Conceptualising how people 'talk' and 'do' generations in contemporary society. *Advances in Life Course Research*, 24, 1–9.

Tornstam, L. (1992). The quo vadis of gerontology: On the scientific paradigm of gerontology. *The Gerontologist*, 32(3), 318–326.

Walker, A. (1990). The benefits of old age? Age discrimination and social security. In, E. McEwan (Ed.), *Age: The unrecognised discrimination* (pp. 58–70). London: Age Concern.

Warner, M. E. and Zhang, X. (2019). Planning communities for all ages. *Journal of Planning Education and Research*, 0739456X19828058.

Weiss, D., Job, V., Mathias, M., Grah, S., and Freund, A. M. (2016). The end is (not) near: Ageing, essentialism, and future time perspective. *Developmental Psychology*, 52(6), 996.

Wilson, M. (2016). The Utopian Moment: The language of positivism in modern architecture and urbanism. In, Monteiro, M. D. R., Pereira Neto, M. J., and Kong, M. S. M. (Eds.), *Utopia(s): Worlds and frontiers of the imaginary* (pp. 77–82). London: Routledge.

Wilson, M. (2018). *Moralising space: The Utopian urbanism of the British positivists, 1855–1920*. London: Routledge.

Woodspring, N. (2016). *Baby boomers: Time and ageing bodies*. Bristol: Policy Press.

World Health Organization. (2002). *Active ageing: A policy framework*. Geneva: WHO.

World Health Organization. (2007). *Global age friendly cities: A guide*. Geneva: WHO.

Suggestions for further study

- Hartt, M., and Biglieri, S. (Eds.). (2021). *Aging people, aging places: Experiences, opportunities, and challenges of growing older in Canada*. Bristol: Policy Press.
- CBC Gem. "Never Too Old." CBC Docs Pov Video, 44:07. 22 August 2019. https://gem.cbc.ca/media/cbc-docs-pov/season-3/episode-4/38e815a-0117cdd0cc7.
- Wells, K. "It's Their Life." *Sunday Edition*, the Canadian Broadcasting Corporation. Toronto, ON: CBC, 30 November 2018. https://www.cbc.ca/player/play/2304600412.

6

CHILDREN

Planning playing catch-up

Sukanya Krishnamurthy, Jenny Wood, Teresa Strachan, and Sean Peacock

Introduction

The concept of childhood is widely accepted as a social construct that has changed considerably over time. Often, it is determined by a push and pull between upper-class ideals and economic realities – a pattern that is largely replicated in the birth and evolution of the UK's planning systems since the late 1800s. Anglo society has increasingly perceived children and youth through a polarized lens, which either identifies them as a source of fear or mystery, or as in need of protection from harm. These concerns have been exacerbated by the environments that children and young people (CYP) inhabit (Horschelmann and van Blerk 2013).

In recent decades awareness has increased at the supranational stage on the role of the built environment in children's health and well-being. A push coming from researchers in the fields of public health, children's development, planning, and geography, has instigated a wave of research and policy responses. These range from UNICEF's Child Friendly Cities initiative to the UNESCO program "Growing Up in Cities." It is now widely acknowledged that children experience cities differently from adults, and their needs in urban space often differ from adult planning prescriptions, even where these are thought to be in the best interests of children (Chatterjee 2015). However, the extent to which this knowledge and understanding has translated into child-friendly planning approaches and environments is questionable. Gillespie (2013, 75) notes that "children's dependence and need for protection, segregation, and delayed responsibility are culturally constructed norms that have served to shape urban planning." But within planning practice children have historically had few or no independently recognized rights.

Much has been written about the importance of play and associated services (Karsten 2005; Krishnamurthy 2019, 2020), access to nature and green

DOI: 10.4324/9781003157588-7

space (Chawla 2015), and children's rights (Tisdall 2008). This chapter discusses the evolution of child-friendly planning in the United Kingdom and the core national and international drivers since the early 20th century. The chapter also explores contemporary opportunities and challenges, including the United Nations Convention on the Rights of the Child (1989), the UN's Sustainable Development Goals, digital technologies, and the global reach of planning with and for children. If we are to plan for more equitable cities, we must embrace the values underpinning planning for child-friendly environments, and understand culture, context, history, and existing policies and funding mechanisms.

Late 1800s–1945: planning for families

Current thought frames childhood as a time of discovery, innocence, and learning, with set standards and milestones of education and maturation. Many adults nostalgically reminisce about hours spent playing and holding naïve beliefs in their youth. However, there was a time just over a century ago when childhood was spent working in jobs which were considered then as ideally suited to children's intellect and size. Before the 1870s, the UK children frequently worked in factories. Only in 1878 was child labor eliminated for those under ten (based on the Factory Act of 1878). Soon afterwards, in 1880, schooling for children between the ages of five and ten become compulsory for the first time (Education Act 1880). At that time, there was very little formal consideration of children's and families' needs in urban space and housing.

Progress was made for working class families as housing reforms of the 1890s, social reforms of the early 20th century, and post-World War I prosperity led to slum clearances and a mass home-building program throughout the United Kingdom. This improved living conditions and life prospects for many (Hall 1975). The public health issues associated with overcrowding and slum housing in a rapidly urbanizing country provided a key push for a nationalized planning system, which was formalized in 1947. The Garden City movement is an early example where social ideals and opportunities for quality local services were considered. However, few Garden City prototypes were actually built.

Alongside developments in urban planning was a growing consciousness of children's welfare, which had been emerging since the 1920s. The culmination of this was the 1924 Declaration of the Rights of the Child, drafted by British social reformer Eglantyne Jebb to recognize that society "owes to the Child the best that it has to give" (OHCHR 2007, 28). Signed by the League of Nations, this declaration paved the way for subsequent conventions that led to a global recognition of the need for more child-friendly environments in recent decades.

Unfortunately, the mass public-sector home-building efforts of the inter-war period were not always accompanied by equitable allocation of community facilities, schools, and services. The results were long commuting distances and decreased access to daily services for many children and families (Clapson 1998). It was not until the social reforms of the post-World War II era and the development of the 1947 Town and Country Planning Act that the issue of accessibility started to be addressed. Some post-World War II reforms stemmed from the evacuation of children from inner-city areas during World War II to rural areas, which unveiled the disparity in the health and well-being of children in rich and poor households. This awareness contributed to the UK's postwar focus on improving children's lives and was instrumental in molding the modern welfare state (Cunningham 2005).

1945–1960: planning without children

The decentralization of communities from major urban centers was a cornerstone of postwar planning. In the United Kingdom, it was spearheaded by the New Towns Act of 1946, which provided new opportunities and funding to fully masterplan communities. A program of building New Towns ran for 24 years, creating 32 towns in total (DCLG 2006). These were envisioned as "balanced and self-contained communities for working and living" (New Towns Committee 1946, 2). Like the Garden City, the New Towns adopted the "third way" between urban and countryside. The ideals and intentions advanced by this program improved the living conditions of many working-class families. For instance, the first New Town (Stevenage) pioneered the idea of a fully pedestrianized town center in the United Kingdom. Notably, early New Towns acquired the nickname "pram towns," owing to their appeal among families with children (Alexander 2009).

The need for rebuilding cities after World War II brought about rapid redevelopment with spaces better organized to promote the health and well-being of citizens. This led to both progress and regression for the lives of children and families. A program of urban renewal replaced cramped and unsanitary slum dwellings in industrial cities with new, more spacious housing, in which children could have their own rooms. Alongside new housing, local authorities allocated land for open space, parks, and schools. The 1950s and 1960s saw new council-built housing estates spring up on urban edges to cater to a growing population and replace ageing homes in inner-city areas.

Yet, the post-World War II era also created disadvantages for children and their families. In seeking to meet the material needs and wants of families, social, and psychological prerequisites were arguably neglected. As the UK cities grew, children's freedoms began to shrink. Ward (1990, 19) describes this as a process where children were "caught in a cage where there is not even an illusion of freedom of action to change [their] situation." He lamented that "self-confidence and purposeful self-respect drain away from these children as they grow up because there is no way which makes sense to them." Although many societal factors contributed to children's shrinking freedoms, planning decisions in the 1950s and 1960s played a significant role.

Suburban estates, especially high-rise ones, were a mixed blessing for children, and well-intentioned New Towns did not always live up to the utopian ideals of their forbearers. For instance, a modernist experiment in recreating "streets in the sky" forced some young city dwellers to live and play 20 or 30 floors up from the ground. Some high-rise buildings and estates did integrate elevated playgrounds or dedicated play spaces at ground level but these provisions were often unsuited for children's informal play (Wright et al. 2019). Similarly, in low-rise suburban estates playgrounds were often an afterthought and suffered from poor maintenance (Wheway and Millward 1997). Ultimately, "streets in the sky" and formalized playgrounds represented poor alternatives to the informal play opportunities created by the low-traffic streets and the green spaces that they replaced (Mackintosh 1982; van Vliet 1982).

Moreover, the location of housing estates on the urban periphery, with limited access to services, contributed to a sense of isolation for families and reduced children's mobility. In turn, commentators now reflected on the impact of decreased informal social contact fueled by rising fears of "stranger danger." Above all, the growth in popularity of the private car undermined children's freedoms immensely. Perversely, modernist planners came to see the compact spatial and functional boundaries created through traditional urban planning as a "quaint and sinister way to decrease personal freedom" (Aldridge 1979, 106). This adult-centric view of freedom and the role of public space often worked to the detriment of children's own independent mobility. It added to the growing opposition to children's street play, with some drawing links between unsupervised play and delinquency, and others seeing it as a reflection of inadequate provision of formal play space (Cowman 2017).

1960–1980: the rise of car-centric planning

By the 1960s, car ownership had become an aspiration among working-class families (Gunn 2011). However, a dramatic rise in car ownership fueled new fears about children's increasing exposure to traffic accidents, and parental anxieties about street play. As early as the mid-1940s, newspapers were reporting that more children died on the roads than adults, with working-class children more likely to be the victims of such accidents (Cowman 2017). By the 1960s, statistics had worsened and many parents started to keep their children away from the streets when unaccompanied.

Movements to create play streets that asserted the right for children, and not cars, to take ownership of urban spaces gained traction too in this era. However, these were the exception and not the rule (Cowman 2017). Indeed, the creation of adventure playgrounds on empty bomb sites in the initial postwar era had enabled some children to navigate and appropriate urban spaces. However, during the 1960s and 1970s these sites were largely reclaimed for other (non-child-focused) developments (Kozlovsky 2008). Conceptions of the playground also became increasingly focused on fixed equipment for structured play. This attended to adult-determined norms of play and served to further segregate the types of spaces children were and were not permitted to occupy.

Car-centric planning in the 1960s brought to light a further issue: council estates built just a decade earlier had not made allowances for the rise in car ownership, meaning that on-street parking became the norm. This dramatically reduced the available space for informal play and intensified fears of child delinquency. Children playing near parked cars came to be seen as a nuisance or a threat that exacerbated a vicious cycle of increasingly polarized views of children and their acceptable conduct. Further planning responses in the 1960s and 1970s did little to combat this.

With an influential government report called Traffic in Towns (Buchanan and Crowther 1963) came a sweeping program of road-building that would segregate transport modes in urban cores and transform existing cities for the motor age. Imposing, elevated highways began to cut through major city centers. While these separated cars from pedestrians, they made cars and car-related development a central feature of city life (Gunn 2011). Highways cut off pedestrian and cycle connections, which had been used by children, and led to dispersed land-use facilities served by motorized transport (Self 1979).

The New Towns were also subject to the changes brought about by the motor age. A complete segregation between pedestrians and vehicles was applied in later New Town designs, along with grid-style layouts and a visible focus on driving. These designs often made non-car journeys longer and forced pedestrians to use walking routes that felt less safe due to poor amenities such as inadequate lighting (Clapson 1998; DCLG 2006). While segregated pedestrian routes were originally designed to promote natural surveillance and informal play, inadequate maintenance led to a perception of feeling unsafe due to a lack of "eyes on the street" (Lock and Ellis 2020).

The well-being of children was the subject of many campaigns against road-building. Jane Jacobs' renowned civic activism in the United States drew attention in the United Kingdom as well, and spotlighted the need to preserve space for incidental play in cities (Jacobs 1961; Ellis et al. 2015). With the social revolution of the 1960s and 1970s came the need to think about how people could be involved in planning and create positive outcomes for and with their communities. Moreover, key scholarly works in the 1970s (e.g., by Colin Ward and Roger Hart) attempted to shed light on the experiences of children in cities and make the case for children's involvement in decisions that affected them. These important works formed the basis for the academic research area now termed children's geographies (Freeman, 2019).

In the policy arena, the UK Government's Skeffington Report (1969) was a landmark in that it examined how people should be more involved in the planning process and shaping planning outcomes. The report showed that there was an appetite for more direct democracy – beyond simply notifying people as to what was happening around them. While the report did not make any specific provisions for including children in urban planning processes, it can be seen as a vital precursor to the later recognition of children's rights to participate in place-making (Wood, 2015).

1980–2010: planning for (and with?) children

By the end of the 20th century, various mechanisms of support had been adopted for planning with and for children in Scandinavia. In this period, the United Kingdom – though not a pioneer in child-friendly planning – drew lessons from other parts of the world. One of the first moves was to ratify the UN's Convention on the Rights of the Child (UNCRC) in 1991 (United Nations 1989). This convention – the most widely ratified human rights treaty in the world – built on historic declarations that mostly attended to children's physical and educational development and protected them from exploitation. The UNCRC represented a step-change towards enshrining children's political and civil rights (a "child" being defined as a person below the age of 18). It provided a framework for bringing about the "three Ps" of *protection, provision*, and *participation*, through 42 interrelated and mutually reinforcing articles. However, its impact on the UK legislation was and remains limited. Instead, commitments to meet the convention were made through policy measures which mostly addressed child services such as education and social work.

Children's participation in placemaking was discussed in a suite of UNCRC articles. These articles attended to children's treatment within the process and the outcomes of decision-making that affected children. The key rights that related to planning were: Article 12: A right to be heard and taken seriously that allows children to express their opinions in all matters affecting them; Article 15: A right to gather and use public space for their own activities, provided they are not breaking the law; and Article 31: A right to play, rest, leisure, and access cultural life.

Building on the momentum generated by the UNCRC, in 1996 UNICEF launched the Child Friendly City initiative (see https://childfriendlycities.org/). This identified the well-being of children as the ultimate indicator of a healthy habitat and a hallmark of a democratic society with good governance (UNICEF 2018a,b). Following up, UNICEF's 2004 document "Building Child Friendly Cities: A Framework for Action" defined a Child Friendly City as "a city, town or community in which the voices, needs, priorities and rights of children are an integral part of public policies, programs and decisions" (UNICEF 2018a,b). More specifically, it was defined as a place where children:

> Are protected from exploitation, violence and abuse; Have a good start in life and grow up healthy and cared for; Have access to quality social services; Experience quality, inclusive and participatory education and skills development; Express their opinions and influence decisions that affect them; Participate in family, cultural, city/community and social life; Live in a safe secure and clean environment with access to green spaces; Meet friends and have places to play and enjoy themselves; Have a fair chance in life regardless of their ethnic origin, religion, income, gender or ability.
> UNICEF (2018a,b, 10)

Apart from supranational level impacts, Planning Aid England and Planning Aid Scotland led local planning initiatives involving CYP through their environmental education programs.[1] Many local authorities were able to draw upon the support of Planning Aid(s) to bring local young people's voices into planning processes. However, since the UK's 2010 general election which replaced the Labour Government with a Conservative and Liberal Democratic coalition, central government funding to Planning Aid England was reduced with any bids to provide community planning advice and neighborhood plan support now placed into a competitive arena. This undermined what English planning authorities could now offer in terms of planning education and support to young people and the wider community. During this time north of the border, Planning Aid Scotland continued to benefit from Scottish Government funding to underpin its inclusive placemaking programs.

2010–present: supranational focus on children and planning

The 2010s brought a renewed, supranational focus on children and planning. Within this, practitioners have sought to move beyond the conception of young people as a homogenous group in society to consider intersectionality, lived experience, and the opportunities open to different young people.

The Equality Act (The UK Government, 2010) made age a protected equality characteristic in the United Kingdom. This requires that public policy prevent unlawful discrimination, pursue ways to further equality between groups, and foster good relations between those sharing a protected characteristic and those that do not. In Scotland, it has been compulsory to produce Equalities Impact Assessments (EQIA) for policy and plans since 2011. Other protected characteristics in the Act are: disability; gender reassignment; marriage and civil partnership; pregnancy and maternity; race; religion or belief; sex; and sexual orientation. These reflect the cross cutting "identities" that position age as only one of a number of considerations in planning and service delivery and critically where physical place takes a key role in shaping those individuals (Hopkins, 2010).

Protected characteristics can all be held by CYP, meaning sensitivity is required in understanding the needs that differ among CYP and how they both conceive of their own rights, and how professionals should understand specific rights in relation to different CYP. For example, the needs of small children and teenagers are not comparable, with the former have differing levels of capability and independence than the latter.

Gender plays a critical role, with adolescent girls' needs less represented in design, planning, and research. A problematic conflation of CYP needs as being the same across all ages and gender can lead both to areas of conflict where children and teens are designated the same space for their leisure (Wood 2016), and a feeling from adults that young people should not be hanging around but engaged in something "productive." Importantly, evidence suggests teenagers, especially girls, are in fact spending less time outdoors than ever. (In a Swedish study, only 20% of teenage users of public space were girls) (Akerman et al. 2017).

The relationship between CYP and their parents and caregivers needs attention too (Krishnamurthy and Ataol 2020). This is particularly so for younger children, whose domain is largely determined by the need to be close to parents and/or caregivers (Williams 2017). The latter also act as gatekeepers to outdoor activity (Veitch et al. 2006). For younger children, outdoor experiences with parents/caregivers, such as playing (Ginsbury 2007), walking (Feigelson 2016), and cycling together (Pooley et al. 2013), are vital for bonding and development. Research has also demonstrated the importance of access to safe spaces and routes (Williams 2017; Krishnamurthy 2019) and community amenities (Thomson et al. 2003) to the mental well-being of parents and caregivers – and consequently to the mental well-being of their dependents.

In addition to a child's age, characteristics such as ethnicity, religion, belief, disability, class, or sexuality mean that the culture of childhood is complex and contextual. For example, Horschelmann and van Blerk (2013) note that LGBTQ2S+ gilds and boys do not have the same opportunity to access networks and spaces to develop their identities as adults, with school and home often representing a "normative" space for heterosexual youth. With respect to ethnicity, these authors also note the structural discrimination and segregation that CYP belonging to an ethnic minority face, drawing attention to higher instances of poverty and social exclusion. In making these arguments, they advocate for a definition of childhood that accounts for intersectionality and heterogeneity and recognizes the individual disadvantage that different CYP face.

The needs of CYPs with disabilities – an estimated 7% of all CYP in the United Kingdom (Blackburn et al. 2010) – are neglected in research. Multiple subgroups exist, with diverse physical, sensory, cognitive, and communication impairments and mental health issues. Each subgroup faces a particular set of barriers to inclusion and participation in urban life (Law and Dunn 1994; Franklin and Sloper 2009; Horton 2017). Features of the environment which might appear trivial to most passers-by, such as availability of transport to and suitable equipment within urban spaces, can constitute fundamental hindrances for CYPs with disabilities.

Alongside movements for child-friendly cities, the UN's New Urban Agenda compiled in 2016 (United Nations 2016), gave CYP the right to participate in all decisions that affected them in relation to where they live. Since the launch of this program, various lessons have been learnt. For instance, planners were made aware that children (and caregivers) the world over expected the urban environment to respond to their needs for: clean safe streets, play spaces, access to green spaces, and adequate services (Brown et al. 2019).

The devolved nations of the United Kingdom are increasingly taking different paths in terms of recognizing children's rights in planning. In England, the key national planning policy documents make little reference either to the need to plan for young people (except in relation to housing) or to specifically engage with them in plan-making (Wood et al. 2019). In Scotland, Planning Aid Scotland continues to receive support of the Scottish Government. Since 2013, Welsh local authorities have been required to consider whether children have sufficient play opportunities in their area, and to prepare an action plan to address any deficiencies, as far as practicable. Emerging evidence suggests that this approach influences the attitudes of decision-makers, and helps raise awareness to children's rights and needs – while it may be too early to assess yet the impact it has on concrete outcomes (Wood 2017a, 2017b). In addition, Scotland and Wales have been taking measures to enshrine the UNCRC in their policy and legislation, to the extent permitted within their respective powers. In Northern Ireland, the devolution settlement is more complex than in Wales and Scotland, and there is a post-conflict legacy to mitigate through planning. Most movement on child-friendly planning has come so far from child-friendly city initiatives established in Derry/Londonderry and Belfast (Wood et al. 2019).

The 2016 UN Sustainable Development Goals (SDGs) provide a potential pathway to improving the lives of children internationally (UNESCO 2016; Britto et al. 2017; Raikes et al. 2017). Going beyond access to basic services for children and youth, the SDGs recognize young people's role in planning for a more equitable and sustainable future. In particular, SDG 11.7, Sustainable Cities and Communities, focuses on providing universal access to safe, inclusive, accessible, and green public spaces by 2030 – especially for women, children, the elderly, and disabled people.

The implementation of the SDGs is expected to play an important role in planning cities with children in mind. Indeed, the SDGs now form part of the Scottish Government's "National Performance Framework" (Scottish Government 2018). This means that any outcomes achieved through Scottish policies and strategies should align with the vision set forth in the SDGs. In Wales, the SDGs have been integrated into a far-reaching legislative instrument called "The Wellbeing of Future Generations (Wales) Act 2015." In England, slower progress has been made given low political interest in this major international initiative. In Northern Ireland, recent and ongoing political disagreement has limited the decision-making capacity of their devolved government.

Since 2017, six cities and communities in the United Kingdom have been working toward official Child Friendly City recognition with UNICEF: Aberdeen (Scotland), Barnet (London), Cardiff (Wales), Derry City and Strabane District (Northern Ireland), Liverpool (England), and Newcastle upon Tyne (England). UNICEF has identified ten optional "badges" to strive for, including "Place," and "Participating," alongside three that are mandatory, "Culture," "Co-operation & Leadership," and "Communication." UNICEF provides a framework that explains what is involved in attaining each of these badges but it also works alongside the local authorities to help them achieve their goals, and eventually conducts an independent review of the results.

One recent project that has aimed to develop a sense of agency among young people is "YES Planning" in Newcastle (Strachan 2018). It focuses on teaching children about their surrounding environment, communities and planning. Forging partnerships with the local planning authority, the urban planning students at Newcastle University lead school-based workshops to explain to children the planning processes that surround new takeaway premises and other local-scale developments. Children are also taught how to role-play a planning committee meeting regarding a hypothetical proposal. This understanding then transitions into a sophisticated reflection on their local neighborhood, how the community uses it and the significance of a responsive planning system in this process.

Conclusion

The planned environment can act as an enabler for CYP to achieve their potential, but this link is complex and multifaceted. Children might feel connected to their local area and empowered through an engagement process that shapes their sense of agency. Yet children's participation in urban planning is still a rare occurrence in the UK context and notable through its absence in planning history and debate. The more children are accepted as "competent social actors" in general (Freeman and Tranter 2012, 224), and therefore citizens capable of contributing to planning and design, the more environments will be created both with and for their needs in mind.

As the agenda of children in cities gains global momentum and visibility, the responsibility of those involved in shaping the urban environment increases. Planners, designers, architects, and other built environment practitioners must acknowledge that childhoods are not universal, nor are children one homogenous group. Needs vary across age, culture, and context, and planning for child-friendly environments needs to respond to these variances. However, planners and designers have limited collective experience in attending to these multifaceted needs.

Now there is a growing movement within the United Kingdom to start engaging more directly with how the needs of CYP differ from those of adults, but also from one another. As societies change, this becomes ever more important. The onset of a digital revolution provides new opportunities for wide engagement, but also changes to how the urban environment is used and perceived. A focus on the well-being of CYP in cities is needed now more than ever, as threats such as the climate emergency, social inequality and injustice, and global pandemics require robust and coordinated responses. The need to move beyond traditional boundaries and embrace holistic approaches with clear prioritization of those who have been historically most unheard and ignored are key going forward.

Alternative planning history and theory timeline

Period	Planning history and theory canon	Alternative timeline: children
Late 1800s–1945	Birth/formalization of planning	*Planning for families* Compulsory schooling, reduction of child labor, relieving overcrowding in cities, housing reform, growing consciousness of children's welfare
1945–1960	Growth of planning	*Planning without children* Recognizing the importance of smaller communities and social facilities, Garden Cities and New Towns program, rapid housebuilding by local authorities, mixed success of planning interventions, shrinking of children's freedoms, rising car ownership
1960–1980	Midlife crisis of planning	*The rise of car-centric planning* Increasing car dominance, increasing fears about children's safety, reduction in street play, recognition that citizens should be involved in planning
1980–2010	Maturation of planning	*Planning for (and with?) children* Securing children's participation rights, Convention of the Rights of the Child, growth of child-friendly city movement
2010–present	New planning crisis	*Supranational focus on children and planning* Funding cuts, mechanisms to support children's participation stripped back, Sustainable Development Goals and new moves to improve children's lives, incorporation of children's rights into law and planning

Case study: children's participation in planning through digital technologies

Digital technologies have transformed how citizens engage with and access public services and spaces. There is little evidence to suggest that children and youth are using technology to formally participate in urban planning, save a handful of demonstrator projects conducted by researchers or practitioners. However, a wealth of global examples show that young people are leveraging the power of social media and the internet to intervene in important local and global causes – such as the climate emergency (Jenkins et al. 2016). The school climate strikes – a movement that was started in August 2018 by Greta Thunberg, a 15-year-old from Sweden, and spread rapidly via social media – are a prominent example. The rise of the digital age is also affecting how childhoods are lived today. For example, the increasing pressures of social media have been linked to increasing rates of mental health difficulties due to phenomena such as "fear of missing out" (FOMO) or less time spent outdoors playing and exercising (Wood and Hamilton 2021). Owing to social media, children and parents are increasingly cognizant of the various dangers present in their local area, and online discussions can exacerbate the fears of the unknown. However, social media can also have benefits for children, by providing services and information that can support identity formation and self-care practices (Cutting and Peacock 2021). Moreover, children's growing digital literacy is encouraging when considered alongside the digital transformation of public participation processes in the future. Many primary school curricula across the United Kingdom now require children to have knowledge of coding and software. Accessible and low-cost tools like the Raspberry Pi and the BBC Micro:bit have helped make this possible, and such tools have been adapted by researchers to involve children in city decisions. For example, Peacock et al. (2020) used a portable sensing kit fashioned out of low-cost equipment to involve primary school children in Newcastle upon Tyne in collecting air pollution and traffic data in their neighborhood. Integrating these data as part of a placemaking process enabled children to generate ideas for the future of their neighborhood and weigh in on local debates about improving their city's environment. Digitalization has also produced significant opportunities to understand, and participate in, city development through the gamification of city building (Mallan et al. 2010). Computational power now allows for the creation of realistic digital twins of cities (Batty 2018), which can hypothetically be used to help children conceptualize the city as a whole, involve children in urban planning decisions, and augment their experiences of public space (Potts et al. 2017; Batty 2018; de Andrade et al. 2020). Adapting games already familiar to some children, such as Minecraft (BlockBuilders 2020) and Cities: Skylines (Brasuell 2020), could help planners reach children directly in their homes, schools, and communities, thus potentially redressing their historical exclusion from urban planning (Peacock et al. 2018).

Note

1 Planning Aid England (PAE) was established as part of the Town and Country Planning Association in 1973 and today operates under its umbrella organization, the Royal Town Planning Institute (RTPI), while retaining its own charitable status. Planning Aid Scotland was set up in the 1990s as is separate entity to the RTPI, and like PAE had its own charitable status, offering support to those not able to pay for planning advice.

References

Akerman, A., Rubin, R., and Lindunger, M., 2017. Planning for equity - insights from the perspective of the girl. In: M. Khan and K. Forrester, eds. Researching with and for Children: Place, Pedagogy and play. Edinburgh: University of Edinburgh, 28.

Aldridge, Meryl. 1979. *British New Towns: A Program without a Policy.* London: Routledge & Kegan Paul.

Alexander, Anthony. 2009. *Britain's New Towns: Garden Cities to Sustainable Communities.* London and New York: Routledge.

Batty, Michael. 2018. 'Digital Twins'. *Environment and Planning B: Urban Analytics and City Science* 45 (5): 817–20.

Blackburn, C.M., Spencer, N.J., and Read, J.M., 2010. Prevalence of childhood disability and the characteristics and circumstances of disabled children in the UK: secondary analysis of the Family Resources Survey. BMC Pediatrics, 10 (1), 21.

BlockBuilders. 2020. 'BlockBuilders'. BlockBuilders. 2020. https://blockbuilders.co.uk.

Brasuell, J. 2020. 'Cities: Skylines as an Urban Planning Tool'. Planetizen – Urban Planning News, Jobs, and Education. https://www.planetizen.com/news/2020/08/110146-cities-skylines-urban-planning-tool.

Britto, Pia R., Stephen J. Lye, Kerrie Proulx, Aisha K. Yousafzai, Stephen G. Matthews, Tyler Vaivada, Rafael Perez-Escamilla, et al. 2017. 'Nurturing Care: Promoting Early Childhood Development'. *The Lancet* 389 (10064): 91–102. https://doi.org/10.1016/S0140-6736(16)31390-3.

Brown, Caroline, Ariane de Lannoy, Deborah McCracken, Tim Gill, Marcus Grant, Hannah Wright, and Samuel Williams. 2019. 'Special Issue: Child-Friendly Cities'. *Cities & Health* 3 (1–2): 1–7.

Buchanan, Colin, and Geoffrey Crowther. 1963. *Traffic in Towns.* London: H.M. Stationery Office.

Chatterjee, Sudeshna. 2015. 'Making Children Matter in Slum Transformations: Lessons from India's National Urban Renewal Mission'. *Journal of Urban Design* 20 (4): 479–506.

Chawla, Louise. 2015. 'Benefits of Nature Contact for Children'. *Journal of Planning Literature* 30 (4): 433–52.

Clapson, Mark. 1998. *Invincible Green Suburbs, Brave New Towns: Social Change and Urban Dispersal in Post-War England.* Manchester: Manchester University Press.

Cowman, Krista. 2017. 'Play Streets: Women, Children and the Problem of Urban Traffic, 1930–1970'. *Social History* 42 (2): 233–56.

Cunningham, Hugh. 2005. *Children and Childhood in Western Society since 1500. Studies in Modern History.* London: Routledge.

Cutting, K., and Peacock, S. (2021), 'Making sense of 'slippages': navigating procedural ethics and doing ethical research with children and young people'. *Children's Geographies.* DOI: 10.1080/14733285.2021.1906404

DCLG. 2006. *Transferable Lessons from the New Towns.* London: DCLG.

de Andrade, Bruno, Alenka Poplin, and Ítalo Sousa de Sena. 2020. 'Minecraft as a Tool for Engaging Children in Urban Planning: A Case Study in Tirol Town, Brazil'. *ISPRS International Journal of Geo-Information* 9 (3): 170.

Ellis, Geraint, Jonna Monaghan, and Laura McDonald. 2015. 'Listening to 'Generation Jacobs': A Case Study in Participatory Engagement for a Child-Friendly City.' *Children, Youth and Environments* 25 (2): 107–27.

Franklin, A. and Sloper, P. 2009. Supporting the Participation of Disabled Children and Young People in Decision-making. *Children & Society*, 23 (1), 3–5.

Freeman, Claire, and Paul J Tranter. 2012. *Children and Their Urban Environment: Changing Worlds.* London: Routledge.

Freeman, Claire. 2019. 'Twenty-Five Years of Children's Geographies: A Planner's Perspective'. *Children's Geographies* 18 (1): 110–121.

Freeman, Claire, Paul Henderson, and Jane Kettle. 1999. *Planning with Children for Better Communities 'The Challenge to Professionals'.* Bristol: The Policy Press.

Gillespie, Judy. 2013. 'Being and Becoming: Writing Children into Planning Theory'. *Planning Theory* 12 (1): 64–80.

Gunn, S. 2011. 'The Buchanan Report, Environment and the Problem of Traffic in 1960s Britain'. *Twentieth Century British History* 22 (4): 521–42.

Hall, Peter. 1975. *Urban & Regional Planning*. Harmondsworth: Penguin Books.

Hopkins, P.E., 2010. *Young People, Place and Identity*. London: Routledge.

Horschelmann, Kathrin, and Lorraine van Blerk. 2013. *Children, Youth and the City*. Abingdon: Taylor & Francis.

Horton, J. 2017. Disabilities, urban natures and children's outdoor play. *Social & Cultural Geography*, 18 (8), 1152–1174.

Jacobs, Jane. 1961. *The Death and Life of Great American Cities*. New York: Random House.

Jenkins, Henry, Sangita Shresthova, Liana Gamber-Thompson, Neta Kligler-Vilenchik, and Arely Zimmerman. 2016. *By Any Media Necessary: The New Youth Activism, Connected Youth and Digital Futures*. New York: NYU Press.

Karsten, Lia. 2005. 'It All Used to Be Better? Different Generations on Continuity and Change in Urban Children's Daily Use of Space'. *Children's Geographies* 3 (3): 275–90.

Kozlovsky, Roy. 2008. 'Adventure Playgrounds and Postwar Reconstruction'. In *Designing Modern Childhoods: History, Space, and the Material Culture of Children*, edited by Marta Gutman and Ning de Coninck-Smith. New Brunswick, NJ: Rutgers University Press.

Krishnamurthy, Sukanya. 2019. 'Reclaiming Spaces: Child Inclusive Urban Design'. *Cities & Health* 3: 86–98. https://doi.org/10.1080/23748834.2019.1586327.

Krishnamurthy, Sukanya. 2020. *A Girl's Day Out: Pune – India*. Bernard van Leer Foundation. Univeristy of Edinburgh.

Krishnamurthy, Sukanya, and Özlemnur Ataol. 2020. *Supporting Urban Childhoods: Observations on Caregiver Use of Public Spaces from Pune (IN) and Istanbul (TR)*. Bernard van Leer Foundation.

Law, M. and Dunn, W. 1994. Perspectives on understanding and changing the environments of children with disabilities. *Physical & Occupational Therapy in Pediatrics*, 13 (3), 1–17.

Lock, Katy, and Hugh Ellis. 2020. *New Towns: The Rise, Fall and Rebirth*. London: RIBA Publishing.

Mackintosh, Elizabeth Ann. 1982. *The Meaning and Effects of Highrise Living for the Middle-Income Family: A Study of Three Highrise Sites in New York City*. New York: City University of New York.

Mallan, Kerry, Marcus Foth, Ruth Greenaway, and Greg T. Young. 2010. 'Serious Playground: Using Second Life to Engage High School Student'. *Learning, Media and Technology* 35 (2): 203–25.

Ministry of Housing Communities and Local Government. 2019. *National Planning Policy Framework*. London: UK Government.

New Towns Committee. 1946. *Final Report (Cmnd 6876)*. London: HMSO.

Newcastle University. n.d. 'Canny Planners Project: A Review of the Workshops for Young People Relating to Local Planning Issues'. https://www.ncl.ac.uk/apl/research/case-studies/canny-planners/.

Office of United Nations High Commissioner for Human Rights (OHCHR). 2007. *Legislative History of the Convention of the Rights of the Child*. Geneva: United Nations. https://www.ohchr.org/Documents/Publications/LegislativeHistorycrc1en.pdf.

Peacock, Sean, Aare Puussaar, and Clara Crivellaro. 2020. 'Sensing Our Streets: Involving Children in Making People-Centred Smart Cities'. In *The Routledge Handbook of Placemaking*, edited by Cara Courage, Tom Borrup, Maria Rosario Jackson, Kylie Legge, Anita Mckeown, Louisa Platt, and Jason Schupbach, 1st ed. Abingdon: Routledge.

Peacock, Sean, Robert Anderson, and Clara Crivellaro. 2018. 'Streets for People: Engaging Children in Placemaking Through a Socio-Technical Process'. In *Proceedings of the 2018 CHI Conference on Human Factors in Computing Systems*, 327:1–14. Montreal, QC, Canada: ACM.

Planning Resource. 2008. 'Planning Aid Empowers City Residents to Plan for Future'. https://www.planningresource.co.uk/article/848785/planning-aid-empowers-city-residents-plan-future?utm_source=website&utm_medium=social.

Pooley, C., Whyatt, D., Walker, M., Davies, G., Coulton, P., and Bamford, W., 2010. Understanding the school journey: integrating data on travel and environment. *Environment and Planning A*, 42 (4), 948–965.

Potts, Ruth, Lisa Jacka, and Lachlan Hartley Yee. 2017. 'Can We 'Catch 'Em All'? An Exploration of the Nexus between Augmented Reality Games, Urban Planning and Urban Design'. *Journal of Urban Design* 22 (6): 866–80.

Raikes, Abbie, Hirokazu Yoshikawa, Pia Rebello Britto, and Iheoma Iruka. 2017. 'Children, Youth and Developmental Science in the 2015-2030 Global Sustainable Development Goals'. *Social Policy Report* 30 (3): 1–23.

Scottish Government. 2018. *Scotland's National Performance Framework Our Purpose, Values and National Outcomes*. Edinburgh: Scottish Government. https://nationalperformance.gov.scot/sites/default/files/documents/NPF%20-%20%20A4%20Booklet%20-%2025_07_2018%20%28002%29.pdf.

Self, Peter. 1979. 'New Towns and the Urban Crisis'. *Town and Country Planning* 47: 4–5.

Skeffington, Arthur. 1969. *People and Planning. Report of the Committee on Public Participation in Planning*. London: HMSO.

Strachan, Teresa Jane. 2018. 'Undercover Placemakers: Transforming the Roles of Young People in Planning'. *Proceedings of the Institution of Civil Engineers-Urban Design and Planning* 171 (1): 5–12.

The UK Government, 2010. *Equality Act 2010*.

Tisdall, E.K.M. 2008. 'Is the Honeymoon Over? Children and Young People's Participation in Public Decision-Making'. *The International Journal of Children's Rights* 16 (3): 419–29.

Thomson, H., Kearns, A., and Petticrew, M. 2003. Assessing the health impact of local amenities: a qualitative study of contrasting experiences of local swimming pool and leisure provision in two areas of Glasgow. *Journal of Epidemiology & Community Health*, 57 (9), 663–667.

UN Committee on the Rights of the Child (UNCRC). 2016. 'Concluding Observations: United Kingdom of Great Britain and Northern Ireland. Consideration of Reports Submitted by States Parties Under Article 44 of the Convention'. 72nd session. United Nations.

UNESCO. 2016. 'Education for People and Planet: Creating Sustainable Futures for All'. *Global Education Monitoring Report*. Paris, France: UNESCO.

UNICEF. 2018a. *Advantage or Paradox? The Challenge for Children and Young People of Growing up Urban*. New York: UNICEF.

UNICEF. 2018b. *UNICEF Child Friendly Cities and Communities Handbook*. New York: UNICEF.

United Nations. 1989. *UN Convention on the Rights of the Child*. Geneva: United Nations.

United Nations. 2016. 'The New Urban Agenda'. A/RES/71/256. Quito: United Nations. http://habitat3.org/wp-content/uploads/NUA-English.pdf.

van Vliet, Willem. 1982. 'Families in Apartment Buildings: Sad Storeys for Children?' Environment and Behavior 15 (2): 211–34.

Veitch, J., Bagley, S., Ball, K., and Salmon, J., 2006. Where do children usually play? A qualitative study of parents' perceptions of influences on children's active free-play. Health & place, 12 (4), 383–393.

Ward, Colin. 1990. The Child in the City. London: Bedford Square Press.

Wheway, Rob, and Alison Millward. 1997. Child's Play: Facilitating Play on Housing Estates. Coventry: Chartered Institute of Housing.

Williams, S. 2017. Cities Alive: Designing for urban childhoods. London: Arup.

Wilson, Alexander, Mark Tewdwr-Jones, and Rob Comber. 2019. 'Urban Planning, Public Participation and Digital Technology: App Development as a Method of Generating Citizen Involvement in Local Planning Processes'. Environment and Planning B: Urban Analytics and City Science 46 (2): 286–302.

Wood, Jenny. 2015. 'To What Extent Does the Scottish Town Planning System Facilitate the UN Convention on the Rights of the Child?' Planning Practice and Research 30 (2): 139–59.

Wood, J. 2016. Space to Participate: children's rights and the Scottish town planning system. PhD Thesis. Heriot-Watt University, Edinburgh.

Wood, Jenny. 2017a. 'Planning for Children's Play: Exploring the 'Forgotten' Right in Welsh and Scottish Policy'. Town Planning Review 88 (5): 579–602.

Wood, Jenny. 2017b. 'What Difference Do Rights Make? Decentering the Governance of Children's Outdoor Play in Scotland and Wales'. In Decentring Urban Governance: Narratives, Resistance and Contestation, edited by Mark Bevir, Kim Mckee, and Peter Matthews, 89–114. Routledge Studies in Governance and Public Policy. Abingdon: Routledge.

Wood, Jenny, Dinah Bornat, and Aude Bicquelet-Lock. 2019. Child Friendly Planning in the UK: A Review. London: Royal Town Planning Institute.

Wood, Jenny, and Jamie Hamilton. 2022. Enabling Independent Travel for Young Scots. Online: A Place in Childhood and Sustrans.

Wright, Valerie, Ade Kearns, Lynn Abram, and Barry Hazley. 2019. 'Planning for Play: Seventy Years of Ineffective Public Policy? The Example of Glasgow, Scotland'. Planning Perspectives 34 (2): 243–63.

Suggestions for further study

- Child Friendly City initiative, UNICEF, https://childfriendlycities.org/.
- Gleeson, B., Sipe, N., eds. 2006. Creating Child Friendly Cities: Reinstating Kids in the City. London and New York: Routledge.
- Horschelmann, K., van Blerk, L. 2013. Children, Youth and the City. Taylor & Francis.
- Gill, T. 2021. Urban Playground: How Child-Friendly Planning and Design can Save Cities. London: RIBA.

7

RELIGIOUS MINORITIES

Planning and Islamophobia

Kevin M. Dunn, Rhonda Itaoui, and Samantha Ngui

Introduction

This chapter reviews international research, from a range of Western societies, which has examined the urban experiences of non-Christian groups. The
authors provide a particular examination of the urban experience of Muslims
since the turn of the 20th century, with an emphasis on encounters with the
planning system (such as the seeking of municipal approval for the development
of places of worship). But the broader issues of religious tolerance and privilege,
as manifest within the city, are the focus of the chapter. These broader issues
include spatial mobility and the ability to develop places of worship. The chapter draws on the experience of Australian cities, Sydney in particular. There
is substantive inequality between Christian and non-Christian groups within
the planning system here. Urban practices occur within a cultural context of
Christian-centrism, and cultural assumptions and stereotypes influence urban
management. Specific planning practices facilitate this influence. The authors
reflect on planning processes and the extent to which they are consistent with
secularism and multiculturalism. The chapter concludes with recommendations
on how urban planners could better grapple with religious diversity and deliver
fairer planning outcomes for diverse religious groups.

1800s–1900: white domination of religiously diverse empires

For more than 40,000 years, the religious landscape of Australia reflected the hundreds of regional cosmologies of Indigenous peoples. In many regions, the actual
morphology of the land was linked to the "Dreaming Stories" that explained creation and deep history. The use of, and access to, land was controlled according
to ancient beliefs. There was "sick country" where development was proscribed,

DOI: 10.4324/9781003157588-8

and there were spaces confined by gender, or by seniority (Jacobs 1993). The crossing of borders between regions required that protocols be met, and there were international trade and other exchanges, such as with the Macassan trepang (sea cucumber) harvesters from Indonesia, who practiced Islam (Macknight 1976). With British colonization these exchanges were curtailed.

The Australian religious landscape was dramatically affected by British colonization. This remains the case so long as there is no formal treaty between Aboriginal people and largely Christian colonizers. This colonization is still celebrated on Australia Day, which annually marks the day (26 January 1788) when the British Navy landed near Sydney and claimed Australia for the British crown. This celebration is an issue for Aboriginal Australians, who refer instead to Invasion Day or Survival Day. The alternative suggestion, in the ongoing "Change the date" debate (www.changethedate.com.au), is that the national Australia Day should not memorialize the invasion, and there are similar debates in North America about Columbus Day.

In the colonial era, Christian missionaries spread across Australia, with a view to converting Aboriginal Australians to Christianity. The land was defined as *terra nullius* (empty land) by the British colonists, which was a profoundly racist eradication of Indigenous Australians, their cultures and beliefs (Anderson 2007). However, there were colonial defenders of Indigenous Australians. Humanists influenced by enlightenment and Christian universalism drew attention to their humanity, presence, and entitlement to human rights (Anderson and Perrin 2007). Christian missionaries were facilitated by the colonial expansion.

It has been argued, and convincingly so, that the Church of England was the established religion in Australia after colonization (Mason 2006:1; Frame 2006). The marking of the Australian religious landscape after British invasion was facilitated by state aid to one religious group – the Church of England (Ngui 2008). Even other Christian denominations were disadvantaged through oppression and persecution (e.g., mostly Irish background Roman Catholics) (Breward 1993). A relationship of privilege existed between the state and Protestant Christian denominations. This privileged relationship benefited Christian groups in establishing Churches and other religious institutions in Sydney, including the granting of land and convict labor, the distribution of state aid through the Clergy and School Lands Corporation, and other forms of financial assistance (Ngui 2008).

The *Church Act, 1836*, disestablished the Church of England as the official faith in Australia, which was in keeping with the disestablishmentarianism that was emerging as consequence of modernist secularization and to assuage sectarianism between Protestants and Catholics. This ostensibly provided legal equality across Christian denominations (Anglicans, Catholics, Presbyterians, Methodists) but not for non-Christians (e.g., Jews who had also come to Australia from Britain). However, the Church of England continued to receive two thirds of all state aid between 1837 and 1841 (Ngui 2008).

1900–1945: white nationalism and Christian-centrism

Australia's path to independent nationhood was through a Federation of states that had formerly been British colonies. In the lead up to that Federation in 1901, the dominant ideology on race began to be influenced by a scientific belief in racial separatism and hierarchy. This aligned well with emerging racist nationalism. A notorious "White Australia" policy was enacted through deportations of non-whites, and various exclusionary border mechanisms (Kamp 2010).

At the point of Federation (1901), there was an official proclamation on the freedom of religion. Section 116 of the *Constitution of Australia* prohibited establishmentarianism – ostensibly to contain sectarian conflict between Catholics and Protestants. While this did curtail Protestant supremacy, it did not challenge Christian privilege. Justice Dixon of the High Court concluded in 1948, that "notwithstanding judicial statements of a contrary tendency, the better opinion appears to be that the Church of England came to New South Wales as the Established Church and that it possessed that status in the colony for some decades" (Justice Dixon, High Court 1948:224). This confounds any claims that there had been a secularist separation of Church and State. Christian domination of the religious landscape of Sydney reflected this privileged relationship. The continued privilege of Christianity is also reflected in the use of Christian prayer in Parliament and at other "secular" public events, and in the dominance of the Christian religious calendar (Fozdar 2011).

Despite this domination, this era did see the emergence of non-Christian communities and temples within Empire cities. These included the so-called "stone mosques" of Perth (completed 1905) and Adelaide (completed 1899) in Australia, built from local quarried rock, and later the London Fazl Mosque (1920s) (Musakhan 1932; Stevens 1989; Naylor and Ryan 2002). They were the bases of community-building for Muslim minorities. Nonetheless, the provisions of the White Australia policy made further family migration and re-union difficult. These racist proscriptions were contrary to a key principle of the British Empire which was for unfettered movement of subjects around the Empire (Cope et al. 1991).

The ideological power behind the White Australia policy began to fade in the 1940s and 1950s. This was associated with the Allied powers' discourses during the Second World War and the development and proclamation of the Universal Declaration on Human Rights (UDHR) in 1948 (Nelson and Dunn 2013). However, racist exclusion and Christian-centrism in Australia persisted as assimilationism.

1945–1965: the "brown empire" strikes back

The Centre for Contemporary Cultural Studies (CCCS) (1982) documented in its landmark text the challenges and profound exclusions that were experienced by "brown Empire subjects" who migrated to Britain in the pre- and post-WWII period. They formed important communities and established religious infrastructure. The CCCS book was evocatively titled *The Empire Strikes Back*, and it documented racist nationalism in Britain.

The postwar immigration program to Australia was intended to increase the population of the country and provide important labor for the booming manufacturing sector (Collins 1988). The mantra of populate or perish espoused by the then Minister for Immigration, Arthur Calwell, was informed by concerns over national security (Calwell 1945:4911): "While the world yearns for peace and abhors war, no one can guarantee that there will be no more war ... Our first requirement is additional population. We need it for reasons of defence and for the fullest expansion of our economy."

While driven by necessity, postwar immigration was characterized by bias. This was embedded in the original design of the program. As early as 1946, Arthur Calwell (1946:508) stated that "It is my hope that for every foreign migrant there will be ten people from the United Kingdom .... Aliens are and will continue to be admitted only in such numbers and of such classes that they can be readily assimilated." Over this period, the Australian Government moved to receive more immigrants from countries other than the United Kingdom. By 1950, almost 200,000 people had arrived in the postwar immigration boom. There was a preference still for British immigrants, as shown in the assisted passage scheme (known as the "£10 Pom"), later extended to Northern and Southern Europeans and also to Turks. However, up until 1958 and the repeal of immigration exclusions, it was still difficult for Turks and non-Europeans to gain entry to Australia.

Postwar migration resulted in greater diversity within the Christian population as well as growth in non-Christian populations. While those identifying as Anglican began to steadily decline from 1947, the number and proportion of Catholics, Pentecostals, and other religious groups began increasing (Bouma 2017). The wave of migration also gave rise to profound changes in religion, and Sydney and Melbourne were crucibles of such change (Dunn and Piracha 2015).

Following the establishment of Australia's first Department of Immigration in 1945, settlement assistance in the form of migrant hostels and language tuition was provided. Such assistance supported the policies of assimilation (Koleth 2010). The assimilation ideals of forsaking heritage in order to meld into the dominant culture also permeated planning practice through the insistence that religious developments align with "local character." Religious minorities adjusted their developments to meet these ideals (Ngui 2008). Applicants for mosques would reduce the planned size of their minarets and domes (or erase them) in order to gain approvals or meet consent conditions.

1965–1980: the uneasy interface between multiculturalism, religious diversity, and secularism

In 1967, the Australian Government signed a migration agreement with the Turkish Government for an orderly process of emigration to Australia. This led to a major migration movement of Muslims to Australia. The number of Turkish-born residents in Australia grew from less than 2,500 in 1966 to more than 100,000 by 1971, and this more than doubled within the following ten years

(Manderson 1988). The outbreak of a civil war in Lebanon in 1975 also propelled a wave of migration to Australia. These "quasi-refugees" were approximately half Muslims and half Christian (Young 1988). Turkish and Lebanese Muslims commenced the difficult task of developing and establishing places of worship and cultural centers. This occurred in the face of community and municipal obstruction (Dunn 2001). During this period, the proportion of the Australian population who were adherents of non-Christian faiths doubled: from less than 1% to 2% by the mid-1980s (Dunn and Piracha 2015).

Starting in the 1970s, religiosity began to decline in terms of official adherence. In the mid-1960s less than 1% said they had no religion; this figure jumped to 11% in 1981, and to 30% in 2016 (Dunn and Piracha 2015). This can be read as an increase in non-belief and perhaps secularism, which is often associated with modernism. During the same period, urban planning also aspired to be secular. This was also the period of the civil rights movement and second-wave feminism in Australia, and it saw the emergence of the concept of multiculturalism. The Australian government ratified UN conventions on prohibiting racism, and on religious freedom and civil and political rights. The passing of the *Racial Discrimination Act* (1975) provided scope for minority groups to protest unfair treatment, although protections for religious minorities were mixed as they are not defined as "racial groups."

1980–2010: cosmopolitanism and post-secularism

After the 1970s, the White Australia policy was in rapid decline, leading to immigration from Asia, and later Africa, bringing non-Christian faiths. Geopolitics and conflict in regions such as Southeast Asia, Lebanon, the Balkans, Iraq, Afghanistan, and Sri Lanka saw many non-Christians flow to Australia. Australia had colonial and geopolitical links to these conflicts, such as links to Afghanistan through the British Empire, and to Vietnam through involvement in that War. These gave rise to the migration of non-Christians (from Indo-China, Iraq, Afghanistan). The Government helped organize and regularize some of these immigration programs (e.g., from Lebanon and Indo-China).

The advent of Cultural Planning and the policy of multiculturalism that emerged in the 1970s provided a new nation-building framework in which Australia's culturally diverse people could live in a "truly just society in which all components [could] enjoy the freedom to make their own distinctive contribution to the family of the nation" (Grassby 1973:15). A core principle of the policy of multiculturalism was that cultural diversity (and the freedom to express culture via language, religion, and so on) could exist within the broader framework of national unity, community harmony, and the maintenance of Australia's liberal-democratic values and laws.

There was an increase, especially from the 1980s, in non-Christian religions and beliefs (Islam, Buddhism, Hinduism, Judaism). But the geography of this effect was uneven (Dunn and Piracha 2015:1652). In Sydney, it was in the inner city where the effects were first seen, with early mosques and Buddhist or Hindu temples developed in the 1980s. At that time, the inner-city suburbs were still run-down, and had not yet gentrified. Jewish migrants (and synagogues) were concentrated in the more affluent eastern suburbs. These developments were frustrated by intolerance, especially when non-Christian groups began developing communities in the sprawling western suburbs where working and middle-classes were building their homes.

The ability of the Australian planning system to deliver substantive religious equality was limited in this period. Religious minorities encountered many barriers when attempting to establish places of worship and private schools. Other nations in the Anglosphere fared no better. A study in the United Kingdom by The Royal Town Planning Institute found that ethnic and racial minorities suffered higher refusal rates for development than Christian proponents (Krishnarayan and Thomas 1993). Refusal rates for South Asian temples in the 1980s in Leicester were double that of "white" applicants indicating the work of race and racism, in addition to religious intolerance (Gale 1999).

Establishing mosques was particularly difficult. In Sydney (Dunn 2001) – as well as several cities in France (Cesari 2005) – almost all developments were met with community opposition and municipal barriers. Planning responses to proposals for mosques were influenced by Islamophobia and stereotypes about Muslims and Islamic fanaticism. These were converted into planning anxieties about noise and parking, as well as anxieties about neighborhood change and cultural takeover (Dunn 2001; Al-Natour 2010). In Montreal, religious development applications were judged against the perceived local need for Islamic religious services (Germain and Gagnon 2003). This exemplifies how narrow definitions of local character and amenity – definitions of what is, and what is not compatible with a local area – impede development proposals from non-Christian groups. A particular challenge emerges when seeking to establish a specific architectural style in an already established cultural landscape (Gale 2004; Ngui 2008).

A planning preoccupation with developments that harmonize with the existing surroundings can have adverse outcomes for Hindu, Muslim, Sikh, and other minority faiths (Peach and Gale 2003). Constructions of cultural misalignment between Muslims and non-Muslims have driven some extreme design injunctions such as the minaret ban in Switzerland, which was established through a national Referendum in 2009, such that the Swiss Constitution now states that "The building of minarets is prohibited" (Article 72, Section 3) (Cheng 2015).

In Sydney, such struggles came to a head in December 2005 with the so-called Cronulla riots. These riots marked one of the most significant race/religion riots in Australian history, whereby Anglo-Australian youth targeted individuals of Middle Eastern background while displaying and voicing racist, anti-immigrant, and anti-Muslim rhetoric. Locals organized a protest publicized through SMS messages that circulated across Sydney. According to the New South Wales police, "more than 270,000 text messages were circulated to incite a 'racially motivated confrontation.'" One text message read, "Just a reminder that Cronulla's 1st wog bashing day is still on this Sunday. Chinks bashing day is on the 27th and the Jews are booked in for early January." Mob violence spilled out onto Cronulla beach, streets, and the local train station, with police and some residents attempting to save victims from the assault of the mob. After the initial violence, efforts began from small numbers of young men of Middle Eastern heritage to retaliate, resulting in arrests on both sides. The riots abated by mid-December but the effects are still felt among Australian Muslims (who constitute less than 3% of the population).

Is Islamophobia (or bias against other religious groups) present within the planning sector as well, and how can it be gauged? A first appropriate measure of religious equality for planning is outcomes. If the religious landscape does not physically reflect the religious composition of society this indicates that there may be a lack of substantive equality. If the identity of an applicant is irrelevant to the merits of a development application then, all other things being equal, development applications for places of worship should have a relatively equal likelihood of gaining approval. Empirical research in Sydney, examining 140 development applications for places of worship in 2000–2006, has revealed substantial religious biases. Christian groups accounted for 80% of applications, 89% of approvals, and 35% of refusals. In stark contrast, non-Christian groups accounted for 20% of applications but only 11% of approvals and 65% of refusals – at a time when more than 94% of all places of worship in Sydney were churches. Clearly, non-Christian groups were less likely to gain approval for their applications. Unequal opportunities to develop places of worship for non-Christian groups are likely due to the fact that these challenge the dominance of the local religious landscape (Ngui 2008).

2010–present: religious hyper-diversity, narrow casting, and extremism

The issue of religious diversity remains unsatisfactorily addressed in the various policies of multiculturalism articulated at the national level. Australia is seen by some as Christian (based on legacy, history, and dominance), by others as secular (based on modernist principles), and by others still as multi-faith (based on multiculturalism policy) (Dunn and Piracha 2015). The latter is the more inclusionary vista, but is complicated by the persistence of inequalities, exclusionary attitudes, and extremism (Koleth 2010, p. 27). Indeed, urban spaces remain racialized (e.g., see Forrest and Dunn 2010; Inwood and Yarbrough 2010; Alderman and Inwood 2016; Allen et al. 2019; Najib and Hopkins 2020). Also, Muslim minorities have come under increased surveillance since the 11 September 2001 terror attacks in New York as Australia joined the "International War on Terrorism" and adopted a national program called Countering Violent Extremism (CVE).

Some cultural and/or religious groups are framed as "in place" and others as "out of place" (Sibley 1995; Gholamhosseini et al. 2019). The geographical dimensions of belonging reflect attitudes toward different cultural and religious groups, with intolerance shaping the geographical distribution of belonging (Forrest and Dunn 2006, 2007) and the ability for religious groups to establish institutions across space. Religious intolerance directed toward religious minorities serves to maintain a hegemonic social, spatial, and political order that excludes non-Christian religious bodies, institutions, and sites from urban spaces (Listerborn 2015). Cumulatively, religious intolerance has geographic effects on how religious minorities navigate and access urban spaces (Itaoui and Dunn 2017; Gholamhosseini et al. 2019; Itaoui 2020; Itaoui et al. 2021).

In Sydney, the exclusionary message of the Cronulla riots has produced ongoing negative impacts on young Muslims' public engagement and citizenship (see Itaoui 2016; Itaoui and Dunn 2017). Islamophobia, as a political construction of who does or does not belong in public spaces, is also responsible for limiting the spatial mobility of young Muslims (Itaoui and Dunn 2017; Itaoui et al. 2021). In several cases, Muslims have reported that they felt excluded from Australian beaches, on the basis that their presence is not permissible (Abdel-Fattah 2017). Such exclusion has been legislated elsewhere, such as in France, where resorts and swimming pools have introduced bans that restrict Muslim women from wearing religious clothing, including burkinis (Muslim-friendly swimwear). Covered worldwide was a case of French police forcing a Muslim woman in a burkini to remove her clothing on a beach in Nice (Quinn 2016).[1]

To describe how intolerance effects urban citizenship, Noble and Poynting (2010) have coined the notion of a spatialized "pedagogy of unbelonging." The "othering" of Australians from migrant and/or religious minority backgrounds can "teach" them to feel less comfortable in certain neighborhoods or even in the wider national space. Anticipation of intolerance produces "inventories of spaces of fear" for religious minorities in public and/or leisure spaces such as beaches, streets, shopping malls, public transport, parks, and sports grounds. Spatial imaginaries of Islamophobia inform Muslims' choices to disengage from geographies of risk (Itaoui and Dunn 2017).

As Muslim Australians are confronted with Islamophobia and biases, the spatial manifestation of Christianity is evolving. Trends in the development of new churches include: (1) declining church attendance; (2) diverse church locations and larger catchment areas than in the past; (3) increases in church sites and facility sizes; (4) emergence of "megachurches"; (5) growth of "niche" churches; (6) increased accessory and auxiliary uses of churches; and (7) online presence of churches (Thomson and Pojani 2019). These trends are significant for the contrast they provide to the development of new non-Christian places of worship. Despite demographic shifts away from dominance by Christian groups, new churches are being developed in a diverse range of settings (industrial, retail centers and school halls) and a recent trend toward larger church buildings has been noted (most evident in the mega church). Declining adherence to Christian faiths has not diminished the value placed on churches (Thomson and Pojani 2019).

Conclusion

Planning academics have argued that achieving substantive equality and inclusive citizenship in multicultural societies requires much more than neutrality (Thompson 2003, 2012). This position is generally consistent with the conclusions reached in critical reviews of secularism for multi-faith western societies (Modood 2005). It is also aligned with the interest in difference, as articulated within the postmodern planning paradigm (Watson and Gibson 1995; Huxley 2000; Sandercock 2000). From this review of alternative planning histories the authors have distilled five key planning practices that can support religious equality in a multi-faith society. Some recommended practices emphasize the competencies of the individual planner whereas others recognize the structural nature of forces like religious intolerance and racism.

First, planning processes ought be "sensitive to cultural difference" (Thompson and Dunn 2002:265). In vocational training parlance this is often described as possessing cultural competencies and may include a working knowledge of the needs and ways of life of the religious groups that are locally present. Planners should also be reflexive on the cultural predilections (privilege) of their own practice, and those of the profession (Thompson 2012). Cultural privilege in the form of Christian-centrism allows obstacles facing non-Christians to remain invisible within urban governance. Moreover, planners should be informed problem-solvers and refer to research strategies that can aid introspection and personal reflexivity (Thompson 2012). Importantly, planners and other urban managers should be as diverse as the citizenry they serve. Cultural capacity training has merit locally for law enforcement, businesses, and educators in order to further cultivate inclusive practices and environments where religious minorities can engage in urban spaces freely and safely.

A second important planning practice to achieve a more equal religious landscape is to consult with religious communities about their needs and desires (Nasser 2004). Substantive equality in a multicultural (and multi-faith) nation requires mechanisms for group representation and participation (Castles 1995). But consultation should extend to outreach.

A third planning practice involves critical reflection on the operations of the local planning system, but also on local social settings. This would help planners and communities better understand how debates about amenity can be negatively used to drive spatial assimilation. It is important to reflect on the challenges and strengths of the local cultural environment. Every locality has a unique mix of intolerance, antipathies, tensions, privilege, and hierarchy (Dunn and McDonald 2001). These relate to the specific cultural mix of each area, the socioeconomic profile, and the history of ethnic relations (Forrest and Dunn 2007). For example, the fortunes of a proposal for a mosque may be heavily influenced by the spread and depth of Islamophobia in a given area. Planning practice needs to be cognizant of this cultural framework and take steps to ameliorate its effects and its presence. Planners and their colleagues can take pre-emptive public relations action if a mosque application is received, with an aim of tempering the possibility and valency of anti-mosque politics and the community relations disasters that those entail. This sort of intervention should draw upon local strengths and capacities (e.g., extant multi-faith or ecumenical networks or key spokespeople) to generate a harmonious development process. Longer term solutions would involve anti-racism education that targets hot spots of racism, such as in schools, as well as in local organizations and workplaces within those urban areas associated with higher levels of identified religious intolerance. These education efforts could be modified according to the perceived level of threat identified by racialized groups in various localities around the city, and the racial literacy of the local community.

Fourth, municipal governments should celebrate the diversity of religious traditions and cultures in their localities by hosting religious festivals, representing diverse religious identities in their local landscape, producing place-branding materials, and promoting the narratives and experiences of diverse members of their community. This will foster an understanding of religious values, and encourage intercultural as well as interfaith dialogue. Activities should work to de-center stereotypes and humanize religious groups within broader communities and provide opportunities for multicultural encounter and cross-cultural contact that encourage bridging within local communities. Four essential conditions are necessary for improving successful intercultural contact: (a) conflicting groups should have contact with one another if the aim is to reduce prevailing intergroup tensions; (b) there should be no competition along group lines within the contact situations; (c) the superordinate intention of the contact should be clear and agreed upon; and (d) relevant institutional authorities must sanction the intergroup contact and endorse a reduction in intergroup tensions (Pederson et al. 2005:23–24). To this end, planning practice should be sensitive to the legacies of past disadvantages, structural discrimination, as well as cultural and language handicaps that recent migrants and new citizens struggle with. In essence, planners should be "campaigners for social justice" (Thompson 2003:290–291). Championing the needs of the less empowered could improve substantive equality across religious groups in a given locality.

A final practice involves outreach to groups who are less empowered when it comes to navigating the planning system. Outreach has been a longstanding component of the access and equity principle of multicultural policy (see Office of Multicultural Affairs 1991). Planners should assist those applicants whose language and cultural skills will put them at a profound disadvantage in the development approvals process. Development applications from more recently arrived religious groups may be weaker due to lower language skills and a poorer grasp of civic processes (Beynon 2007). A substantively equal treatment of each application ought to compensate for discrepancies in the quality of development applications being submitted. To deliver substantive equality the planning system needs to account for the uneven empowerment of applicants. Another manifestation of such outreach should acknowledge and address the fear and peril that religious minorities face. Protection for religious sites and spaces is important. This includes vigilance against forms of vandalism (or even terrorism), as has been seen against Muslim sites across the globe. Religious sites and spaces can be supported in providing security services and technologies to protect their places of gathering and community.

Alternative planning history and theory timeline

Period	Planning history and theory canon	Alternative timeline: religious diversity
Late 1800s–1900	Birth of planning	*White domination of religiously diverse empires* Christian sectarianism, disestablishmentarianism, Christian proselytizing, development of non-Christian communities and buildings in empire cities
1900–1945	Formalization of planning	*White nationalism and Christian-centrism* Immigration Restriction Act 1901, feared external "Other," i.e., "non-white" immigrants, restrictions on movements in/out of Australia, inability to reunite with family, restrictions on citizenship and naturalization rights
1945–1965	Growth of planning	*The "brown empire" strikes back* Post WWII "Populate or Perish" increased migration to Australia, "empire subjects" form non-Christian communities within Europe and settler nations, assisted (European) migration schemes, e.g., British ("£10 Pom") & Northern and Southern Europeans
1965–1980	Midlife crisis of planning	*The uneasy interface between multiculturalism, religious diversity and secularism* Shifts from policies of assimilation (and immigration restriction) to policy of multiculturalism, limited coverage of religion, response to increasing migrant cohorts (particularly from continental Europe), modernist planning and impartiality
1980–2010	Maturation of planning	*Cosmopolitanism and post-secularism* Cultural diversity in Australian cities expanded, multiculturalism provided a new nation-building framework, religious diversity expected to exist within Australia's liberal-democratic framework, increased migration and settlement of diverse communities (including Muslims) across major cities in Australia, development of non-Christian religious sites, advent of Cultural Planning, September 11 terror attacks and the "International War on Terrorism," rights to city and urban citizenship
2010–present	New planning crisis	*Religious hyper-diversity, narrow casting and extremism* Emergence of new denominations, enhanced state use of religious groups for welfare provision, Countering Violent Extremism (CVE) and the surveillance of religious minorities, far-right extremism and the targeting of religious minorities, spatial pedagogies of unbelonging

Case study: celebrating the diversity of Muslim communities in Sydney public spaces

Planning for religious diversity should involve the promotion and celebration of diverse identities and practices in city spaces. This particularly the case for Muslims in localities that are home to large Muslim populations. The holy month of Ramadan in 2022 marked a significant success in the implementation of local festivals that celebrated Islam in public spaces across the globe. In Sydney, an effective strategy employed by Canterbury-Bankstown Council in the western suburbs has been the annual Ramadan street festival on Haldon Street in Lakemba. This month-long night food festival has encouraged the participation of thousands of Sydney siders in Muslim cultures, foods, and practices, promoting intercultural contact and exchange. In 2022, it was estimated that over 1 million people attended the festival, more than half of whom were not residents of the local area. The attendance grew fourfold compared to 2019, with attendees travelling from all parts of Greater Sydney and across the state to participate in the dusk-until-dawn celebration of Muslim cultures. Liverpool City Council in Sydney adopted a similar approach in 2022, hosting the "Most Blessed Nights" Street Food Market every weekend in April. The event marked the sacred month of Ramadan, Passover, Catholic/Protestant Easter, and Orthodox Easter, and celebrated the many cultures and religions that make up the Liverpool region. The City Council noted on its website that the event sought to promote the values of unity, family, and togetherness. The success of this and similar events in delivering effective social planning for Muslims in public spaces is also reflected in examples from the United Kingdom. Eid has been celebrated in London's Trafalgar Square annually for the last 17 years. Thousands of people attend the event, which features live music, children's activities, and food stalls. Similarly, the Ramadan Tent Project in the United Kingdom provides free meals every day of Ramadan in an Open Iftar that welcomes people from all walks of life across various cities in the country. As of 2022, the Open Iftar has convened and connected over 350,000 people across the United Kingdom at some of Britain's most loved and iconic cultural landmarks and institutions. Such cultural festivals, events, and activities collectively demonstrate a positive impact in encouraging cross-cultural exchange, appreciation, and celebration. These are paramount in in multicultural cities. More importantly, festivals symbolically demonstrate an ownership of space among minority religious groups like Muslims, reflecting their belonging, ownership and equal right to the diverse city.

Note

1 In Rennes and Grenoble, the burkini ban in state-run swimming pools was reversed in 2019 and 2022, respectively.

References

Abdel-Fattah, R. (2017). Islamophobia and Australian Muslim political consciousness in the war on terror. *Journal of Intercultural Studies*, 38, 397–411.

Alderman, D.H., and Inwood, J. (2016). Mobility as antiracism work: The "hard driving" of NASCAR's Wendell Scott. *Annals of the American Association of Geographers*, 106(3), 597–611.

Allen, D., Lawhon, M., and Pierce, J. (2019). Placing race: On the resonance of place with black geographies. *Progress in Human Geography*, 43(6), 1001–1019.

Al-Natour, R.J. (2010). Folk devils and the proposed Islamic school in Camden. *Continuum: Journal of Media & Cultural Studies*, 24(4), 573–585.

Anderson, K., & Perrin, C. (2007). 'The miserablest people in the world': race, humanism and the Australian Aborigine. *The Australian Journal of Anthropology*, 18(1), 18–39.

Beynon, D. (2007). *Centres on the Edge: Multicultural Built Environments in Melbourne. Everyday Multiculturalism Conference*. Macquarie University, Centre for Research on Social Inclusion.

Breward, I. (1993). *A History of the Australian Churches*. Sydney, Allen and Unwin.

Bouma, G. and Halafoff, A (2017) "Australia's Changing Religious Profile—Rising Nones and Pentecostals, Declining British Protestants in Superdiversity: Views from the 2016 Census" *Journal for the academic study of religion*, 30(2), 129–143.

Calwell, A. (1945). Commonwealth Parliamentary Debates, House of Representatives, 2 August 1945, Vol. 184, p. 4911.

Calwell, A. (1946). Commonwealth Parliamentary Debates, House of Representatives 22 November 1946, Vol. 189, p. 508.

Castles, S. (1995). Multicultural Citizenship. P. R. Service, Parliamentary Research Service: 32.

Centre for Contemporary Cultural Studies. (1982). *The Empire Strikes Back: Race and Racism in 70s Britain*. London: Hutchinson.

Cesari, J. (2005). Mosques in French cities: Towards the end of a conflict? *Journal of Ethnic and Migration Studies*, 31(6), 1025–1043.

Cheng, J. (2015). Islamophobia, Muslimophobia or racism? Parliamentary discourses on Islam and Muslims in debates on the minaret ban in Switzerland. *Discourse and Society*, 26(5), 562–586.

Collins, J. (1988). *Migrant Hands in a Distant Land: Australia's Post-War Immigration*. London: Pluto Press.

Cope, B., Castles, S., and Kalantzis, M. (1991). *Immigration, Ethnic Conflicts and Social Cohesion*. Bureau of Immigration Research, Australian Government Publishing Service, Canberra.

Dunn, Kevin M. (2001). Representations of Islam in the politics of mosque development in Sydney. *Tijdschrift voor Economische en Sociale Geografie*, 92(3), 291–308.

Dunn, Kevin M., and McDonald, A. (2001). The geography of racisms in NSW: A theoretical exploration and some preliminary findings. *Australian Geographer*, 32(1), 29–44.

Dunn, Kevin M., and Piracha A. (2015). The Multifaith City in an era of post-secularism: The complicated geographies of Christians, non-Christians and non-faithful across Sydney, Australia. In S.D. Brunn (Ed.), *The Changing World Religion Map* (Chapter 86, pp. 1635–1654). New York: Springer.

Forrest, J., and Dunn, K. (2006). Racism and intolerance in Eastern Australia: A geographic perspective. *Australian Geographer*, 37, 167–186.

Forrest, J., and Dunn, K. (2007). Constructing racism in Sydney, Australia's largest EthniCity. *Urban Studies*, 44, 699–721.

Forrest, J., and Dunn, K. (2010). Attitudes to multicultural values in diverse spaces in Australia's immigrant cities, Sydney and Melbourne. *Space and Polity*, 14, 81–102.

Fozdar, F. (2011). The 'Choirboy' and the 'Mad Monk': Christianity, Islam, Australia's political landscape and prospects for multiculturalism. *Journal of intercultural studies*, 32(6), 621–636.

Frame, T. (2006). *Church and State: Australia's Imaginary Wall*. Sydney: UNSW Press.

Gale, R. (1999). Pride of place and places: South Asian religious groups and the city planning authority in Leicester. School of Oriental and African Studies. London, University of London. MA Thesis.

Gale, R. (2004). The multicultural city and the politics of religious architecture: Urban planning, mosques and meaning-making in Birmingham, UK. *Built Environment*, 30(1), 30–44.

Germain, A., and Gagnon, J.E. (2003). Minority places of worship in Montreal. *Planning Theory and Practice*, 4(3), 295.

Gholamhosseini, R., Pojani, D., Mateo Babiano, I., Johnson, L., and Minnery, J. (2019). The place of public space in the lives of Middle Eastern women migrants in Australia. *Journal of Urban Design*, 24(2), 269–289.

Grassby, A. (1973) *A Multi-Cultural Society for the Future*, Department of Immigration, Australian Government Publishing Service, Canberra.

Huxley, M. (2000). The limits to communicative planning. *Journal of Planning Education and Research*, 19, 369–377.

Inwood, J.F., Alderman, D., and Williams, J. (2015). "Where Do We Go From Here?" Transportation justice and the struggle for equal access. *Southeastern Geographer*, 55, 417–433.

Itaoui, R. (2016). The geography of Islamophobia in Sydney: Mapping the spatial imaginaries of young Muslims. *Australian Geographer*, 47(3), 261–279.

Itaoui, R. (2020). Mapping perceptions of Islamophobia in the San Francisco Bay Area, California. *Social & Cultural Geography*, 21(4), 479–506.

Itaoui, R., Dufty-Jones, R., and Dunn, K.M. (2021). Anti-racism Muslim mobilities in the San Francisco Bay Area. *Mobilities*, *16*(6), 888–904.

Itaoui, R., and Dunn, K.M. (2017). Media representations of racism and spatial mobility: Young Muslim (un)belonging in a post-Cronulla riot Sutherland. *Journal of Intercultural Studies*, 38(3), 315–332.

Jacobs, J. (1993) 'Shake 'im this country': the mapping of the Aboriginal sacred in Australia. In P. Jackson, & J. Penrose (Eds.), *Constructions of Race, Place and Nation* (pp. 100–118). UCL Press.

Kamp, A. (2010). Formative geographies of belonging in White Australia: constructing the national self and other in parliamentary debate, 1901. *Geographical research*, 48(4), 411–426.

Krishnarayan, V., and Thomas, H. (1993). *Ethnic Minorities and the Planning System*. London: Royal Town Planning Institute.

Listerborn, C. (2015). Geographies of the veil: Violent encounters in urban public spaces in Malmö, Sweden. *Social & Cultural Geography*, 16(1), 95–115.

Macknight, C.C. (1976) *The Voyage to Marege': Macassan Trepangers in Northern Australia*, Melbourne University Press, Melbourne.

Manderson, L. (1988). Turks. In J. Jupp (Ed.), *The Australian People* (pp. 818–825). Sydney: Angus & Robertson.

Mason, J.K. (2006). *Law and Religion in Australia*. Canberra: National Forum on Australia's Christian Heritage.

Modood, T. (2005). *Racism, Ethnicity and Muslims in Britain Minneapolis*. Minneapolis: University of Minnesota Press.

Musakhan, M.H. (ed.) (1932). *Islam in Australia: 1863–1932*. Adelaide: Mahomet Allum.

Najib, K., and Hopkins, P. (2020). Spaces of Islamophobia and spaces of inequality in Greater Paris. *Environment and Planning C: Politics and Space*, 0 (0), 2–20.

Nasser, N. (2004). From multicultural urbanities to the postmodern cosmopolis: A praxis for urban democracy. *Built Environment*, 30(1), 16.

Naylor, S., and Ryan, J.R. (2002). The mosque in the suburbs: Negotiating religion and ethnicity in South London. *Social and Cultural Geography*, 3(1), 39–59.

Nelson, J., and Dunn, Kevin M. (2013). Racism and anti-racism. In A. Jakubowicz, and C. Ho (Eds.), *"For those who've come across the seas"…: Australian Multicultural Theory Policy and Practice* (Chapter 22, pp. 259–276). North Melbourne, Vic: Australian Scholarly Publishing.

Ngui, S. (2008). Freedom to worship: Frameworks for the realisation of religious minority rights, PhD Thesis, University of New South Wales, Sydney.

Noble, G., and Poynting, S. (2010). White lines: The intercultural politics of everyday movement in social spaces. *Journal of Intercultural Studies*, 31(5), 489–505.

Office of Multicultural Affairs. (1991). *Making It Happen: Access and Equity at Work around Australia*. Canberra: Australian Government Publishing Service.

Peach, C., and Gale, R. (2003). Muslims, Hindus, and Sikhs in the new religious landscape of England. *The Geographical Review*, 93(4), 469.

Quinn, B. (2016). French police make woman remove clothing on Nice beach following burkini ban. *The Guardian* [Online]: https://www.theguardian.com/world/2016/aug/24/french-police-make-woman-remove-burkini-on-nice-beach.

Sandercock, L. (2000). When strangers become neighbours: Managing cities of difference. *Planning Theory & Practice*, 1(1), 13–30.

Sibley, D. (1995). *Geographies of Exclusion*. London: Routledge.

Stevens, C. (1989). *Tin Mosques & Ghantowns: A History of Afghan Cameldrivers in Australia*. Melbourne: Oxford University Press.

Thompson, S. (2003). Planning and multiculturalism: A reflection on Australian local practice. *Planning Theory & Practice*, 4(3), 275–293.

Thompson, S. (2012). Planning for diverse communities. In S. Thompson, and P. Maginn (Eds.), *Planning Australia: An Overview of Urban and Regional Planning* (2nd ed.). Cambridge: Cambridge University Press, 226–253.

Thompson, S., and Dunn, K.M. (2002). Multicultural services in local government in Australia: An uneven tale of access and equity. *Urban Policy and Research*, 20(3), 263–279.

Thomson, M., and Pojani, D. (2019). *Emerging Trends of Modern Churches and Spatial Planning Implications. UQ|UP Research Paper no. 1* (pp. 1–15). Australia: The University of Queensland.

Watson, S., and Gibson, K.D. (1995). Postmodern politics and planning: A postscript. In S. Watson and K.D. Gibson (Eds.), *Postmodern Cities and Spaces* (pp. 254–264). Oxford: Blackwell.

Winchester, H., and O'Neil., P. (1992). Semantic differential scales: An idea for urban fieldwork. *Geography Bulletin*, 24, 113–116.

Young, C.M. (1988). Lebanese migration since 1970. In J. Jupp (Ed.), *The Australian People* (pp. 672–677). Sydney: Angus & Robertson.

Suggestions for further study

- Abdel-Fattah, R. 2021. *Coming of Age in the War on Terror*. Sydney: New South Publishing.
- Najib, K., 2021. *Spatialized Islamophobia*. London: Routledge.
- Fahmi, A. 2015. The Cronulla riots made my world a scarier place. *Sydney Morning Herald* 12 December. Available at: https://www.smh.com.au/opinion/asma-fahmi-remembers-cronulla-riots-10-years-on-20151212-glm6c3.html.
- The Conversation, 2016. Rhonda Itaoui on navigating the city as a young Muslim. [podcast] *Speaking with*. Available at: https://theconversation.com/speaking-with-rhonda-itaoui-on-navigating-the-city-as-a-young-muslim-53166.
- SBS Australia. 2015. The Day that Shocked the Nation: Cronulla riots documentary. Available at: https://www.sbs.com.au/cronullariots/documentary.html.

8

PEOPLE OF COLOR

Confronting white supremacy in planning

Melissa Heil, Lou Turner, and David Wilson

Introduction

A gaping contradiction exists in the practices of urban planning when viewed through the lens of Black[1] urban history and politics. A field established to improve cities for people's well-being has, throughout its history, harmed Black communities (Rutland 2018). The popular narrative that urban planning produces more humane, livable cities breaks down when the experiences of Black communities and their residents are centered. As former Director of Research for the Chicago Urban League, Harold Baron (1968, 8), observed:

> … all [of planning's] professional sophistication seems far removed from the number one problem of the city, racism. The aware black man in the ghetto tends to view urban planning more as an enemy than as an aid. There should be no wonder in this, for public housing monstrosities and urban renewal, with its callous treatment in the displacement of Negro householders, are the two most obvious and familiar evidences of planning to him.

This chapter aims to excavate the historical layers of racial inequality produced through urban planning policies and programs in America that continue to shape contemporary racial disparities in wealth, health, and opportunity.

The core of the dilemma: market economies and market-oriented cities, ever unstable and vulnerable, have required planning support that has not been neutral or value-free (Stein 2019).[2] Over time, planners have produced the spaces and spatial fixes needed to meet capitalism's changing needs. Ideas of race have been central to planners' involvement in such economic processes, powerfully embedded in decision-makers' minds throughout planning history

DOI: 10.4324/9781003157588-9

(D. Wilson 2018). As planners have responded to capitalism's shifts (the continuous remaking of land and labor markets, fluctuations in global spatial divisions of labor, etc.), their actions – informed by hierarchical racial thinking – have advanced racialized uneven development in urban America.

In this context, urban planning throughout its history has not positioned Black people as urban citizens whose well-being should be served, but rather as pathologized populations, a problem for planners to solve through containment, management, and expulsion (Drake and Cayton 1970; Rutland 2018). While Blackness has long been viewed as aberrant, whiteness has been viewed as normative (Goetz et al. 2020). The effort to improve the city for human well-being (rhetorically articulated as universally beneficial) has often meant structuring the city for white communities to accrue social, economic, and political advantages (Roithmayr 2010). In this way, urban planning has contributed significantly to structural racism, defined below (Taylor Jr and Cole 2001, 5):

> A distributive system that determines the possibilities and constraints within which people of color are forced to act. The system involves the operation of racialized structural relationships that produce the unequal distribution of material resources, such as jobs, income, housing, neighborhood conditions, and access to opportunities ... The operation of these racialized structures fuses together social class inequality and racial inequality. A characteristic feature of structural racism and inequality is its ability to hide, camouflage, disguise, and conceal its true nature, which causes it to be an insidious force ... that perpetually produces race and social class inequality ...

This chapter is guided by the Black geographies tradition which recognizes the simultaneity of historically entrenched racism in American urban planning and Black communities' continuous struggles against these measures (McKittrick and Woods 2007; Hawthorne 2019). A significant aspect of this struggle has been the production of spaces and spatial imaginaries to enable better lives. Black communities have been active co-producers of the city, making spaces where Black lives matter, working to overthrow racist structures that continue to produce inequitable cities. As such, throughout this chapter, we also chronicle examples of how Black people's resistance has altered practices of city-building and the field of urban planning.

Late 1800s–1900: the Jim Crow era and planning

The field of urban planning traces its origins to the late 1800s, in the heart of the Jim Crow (legalized racial segregation in public facilities and transportation) era, when industrialization brought population booms to large the U.S. cities that created overcrowding and poor sanitation. Disease outbreaks were frequent and deadly. In this setting, urban reformers guided by planning measures sought to improve public health by transforming the built environment (e.g., better sanitation systems, preservation of open space) (Peterson 2003). The poor quickly occupied a central place in the urban planning narrative: many urban reformers believed the physical environment's transformation could also improve this group's moral character and thereby overcome the vices keeping people in poverty (Von Hoffman 1998). Their efforts inaugurated an assumption that persists to this day: that poverty, and especially Black poverty, was exacerbated by the decay of the physical-environmental. Deterioration of the built environment, by this logic, led to the poor's moral and cultural erosion (Teaford 1990).

Sociologist W.E.B. Du Bois was the first to study the conditions of Black communities in the industrial city. In contrast to other urban reformers, Du Bois argued that the value of altering the built environment was limited if broader structures of racism remained unaddressed. His analysis revealed that discrimination – rather than the built environment and individual moral character – afflicted African American communities and confined their residents to dangerous and unhealthy living conditions. Employment for Black city dwellers was limited primarily to service work and general labor, shaped by racist ideas about the kinds of work to which they were "naturally" suited. This resulted in their exclusion from many jobs, lower wages, and frequently unstable employment realities (Du Bois 1899).

Working primarily in service positions, many Black residents needed to live near their employers – wealthy households, hotels, and stores – to retain employment. At the same time, housing access was deeply constrained by discrimination. Most landlords would not rent to African Americans, and there were many incidents of violence directed at Black families who sought to settle in predominately white areas (J. Trotter, Lewis, and Hunter 2004). Because housing options were limited, landlords could charge a premium for substandard housing even as employment discrimination meant Black households had lower incomes than other groups (Du Bois 1899). Given these limited housing options and their outsized cost, many African Americans lived in overcrowded conditions (sub-renting rooms to lodgers) in structures that lacked adequate sanitation and ventilation.

As a result of such crowded living conditions, Black city dwellers were disproportionately affected by disease outbreaks. For example, Black New Yorkers and Atlantans died from tuberculosis and pneumonia at significantly higher rates than whites (United States Bureau of Labor 1897; Osofsky 1971). Poverty and discrimination by medical professionals reduced Black communities' access to treatment and contributed to higher fatality rates. These disparities were often blamed on supposed "inherent racial inferiority," fueling beliefs that African-American neighborhoods were disease vectors that threatened public health (Schlabach 2019).[3]

This period planted the seeds of ideas that would define the relationship between urban planners and African American communities for decades. Poverty was presumed to result from low moral character and inferior environmental conditions rather than human-discriminatory and human-punishing practices. Vulnerability to disease, attributed to inherent biological otherness, cast Black communities as dangerous nodes in cities – a threat to be contained. Together these ideas set the groundwork for racially punitive urban policies to unfold (Teaford 1990).

Yet, in the face of discrimination in this period, African Americans organized to build Black spaces of nurture, resilience, and resistance. In Black neighborhoods, civil society organizations and charitable associations formed to meet the community's needs and advocate for improved conditions for people of color in urban society (Du Bois 1899). Similarly, Black freedom towns were established throughout the country, built by and for emancipated freedmen. These towns were built away from white populations with the intent to create spaces of Black self-determination (local governance that was separated from the white supremacist power structures of existing communities) and Black cultural enrichment (through the creation of educational, civic, and religious organizations) (Cha-Jua 2000; Corn et al. 2016). Black freedom towns and urban neighborhoods – while always contending with the structural constraints of racism and outright violence – were spaces that made new ways of life viable for Black people and would become cradles for liberation organizing over the next century.

1900–1945: urban planning builds the ghetto

The urban planning profession expanded and deepened its sphere of influence between 1900 and 1945. Two new developments in the period set the stage for its systematic creating of Black ghettos as its most searing legacy: the formalization of the planning profession and the beginning of the Great Migration (the 60-year mass migration of Black Americans from the rural South to Northern and Western cities). First, planning became widely institutionalized as a profession and discipline in American universities, growing as a powerful formal presence in city governments (Teaford 1990). Second, Black populations in American cities grew dramatically. Pushed out of the South by disruptions in the agricultural economy and Jim Crow, Black Americans sought new opportunities with expanding industries in urban centers (J. W. Trotter 1991). With over 6 million African Americans relocating between 1910 and 1970, urban African Americans – disproportionately working poor and stigmatized – became people to be assiduously managed by planners (Shabazz 2015).

The rise of new, legally sanctioned planning tools followed. These tools, notably zoning, were immediately put to work to manage Black communities. This early-20th-century innovation enabled state planners, for the first time, to assert control over private property to guide the development of cities. Zoning, ostensibly operating to stabilize land values and promote land uses conducive to people's well-being in the city, also functioned as a race-class exclusionary mechanism (Silver 1997). In 1910, Baltimore enacted the country's first racial zoning ordinance, which legally restricted where African American residents could live, keeping them spatially isolated in blocks that were already majority Black (Farrar 1998). Baltimore's mayor argued that this would protect white neighborhoods from the spread of disease and preserve property values. Black residents immediately began organizing against the effort, helping to lead what would become a global movement against legal segregation (Nightingale 2012). Local governments in mid-Atlantic and southern cities followed Baltimore in adopting racial zoning ordinances until the practice was declared unconstitutional by the Supreme Court in 1917 (Silver 1997).

Even after racial zoning was declared illegal, zoning remained a critical tool for enforcing segregation, relying on planners to provide technical information and rationales that could be used to enact segregation plans in legally defensible ways without direct reference to race (Silver 1997). Planners began developing exclusionary zoning guidance based on class (e.g., forbidding multiple-family dwellings) to achieve the same goal of racial exclusion, making many neighborhoods financially inaccessible for most Black city dwellers (Ritzdorf 1997). As a result, zoning could still structure cities according to apartheid logics. Furthermore, zoning was used to undermine the stability of Black neighborhoods, allowing industrial uses in or near Black residential areas. This disproportionately exposed African American residents to environmental hazards. These neighborhoods were also regularly zoned as vice districts, allowing commercial uses not permitted in other parts of the city, such as gambling houses. Such uses deepened the popular association of Black neighborhoods with criminality (Rabin 1989). In these ways, zoning made Black neighborhoods less desirable, less safe, and less healthy.

In 1934, federal policy introduced another mechanism that would deepen segregation and the wealth gap between white and Black Americans: redlining. In the 1934 Housing Act, the Federal Housing Administration and the Home Owners Loan Corporation were empowered to insure mortgages and extend new credit lines to potential homebuyers, making lending more widely available. Lending practices were guided by "residential security maps" that divided cities into zones based on how risky real-estate investments were assessed to be. Mortgages in areas that were deemed to be high risk – marked in red on lenders' maps – would not be eligible to be insured. Among the factors for designating an area high-risk was the presence of people of color who were considered a risk to housing values (Jackson 1987). Across the country, Home Owners Loan Corporation documents described Black, Latino, and Asian (as well as Jewish, Italian, and Eastern European) populations as "undesirable," "subversive," and "lower grade" infiltrators whose presence was likely to depress property values ("Mapping Inequality: Redlining in New Deal America" 2016). This made it more difficult for Black Americans to access credit to purchase homes in the limited areas they were permitted to live, preventing many Black households from building wealth through property ownership. Those who sought to buy often had to rely on the riskier method of "contract buying," in which ownership of the property was not transferred to the buyer until the house's total price had been paid. Missing one monthly payment could cause a buyer to lose their entire investment.[4] In Chicago, contract buying is estimated to have cost Black communities at least $3.2 billion in additional housing expenses (George et al. 2019).

1945–1965: planning through demolition

The postwar era saw urban planners responding to the rollout of a major new development: significant suburban development. With the GI Bill and the National Highway Defense Act of 1956 fueling suburbanization,[5] city centers were to be stabilized with the aid of the new federal urban renewal program. Urban renewal was launched as part of the Housing Act of 1949[6] as a mechanism to clear slums and invigorate new housing construction, which had slowed considerably during the Great Depression and World War II. The program's emphasis on slum clearance was supposed to allow for easy assemblage of land using eminent domain (government expropriation of land for public re-use). Government-funded demolition would jumpstart redevelopment since neither current owners nor future investors were willing to take on the cost of demolishing existing structures (Weiss 1980). While the program dramatically redeveloped neighborhoods throughout the country, it failed to improve housing conditions for many. Nationally, the urban renewal program demolished 600,000 housing units and displaced 2 million people (Talen 2014; Levy 2016). In slum clearance locations, only 250,000 new housing units were constructed. Instead, many residential areas were converted to commercial spaces, often for hospitals, higher education institutions, and shopping districts.

Urban renewal represented a massive investment in urban redevelopment ($13 billion between 1949 and the program's end in 1973) ("Renewing Inequality" 2016), but in many instances deepened racial inequality. Local and national policymakers often viewed communities of color as threats to cities' economic and social vitality (Rothstein 2017). They became targets for demolition. Through urban renewal and new highway construction (see Bullard, Johnson, and Torres 2004; Connolly 2014), cities across the country tore down centers of Black social and economic life. This destruction led to material dispossession, psychological trauma, and collapsed social and political networks in African American communities (Fullilove 2016). In Washington DC, Baltimore, Detroit, St. Louis, Atlanta, and Philadelphia, people of color comprised two-thirds (or more) of people displaced by urban renewal projects ("Renewing Inequality" 2016). Across America, urban renewal demolished Black neighborhoods (deemed blighted, overcrowded, and undesirable) and remade these city spaces for white consumption. The program ultimately targeted one racial group for exclusion and expulsion, while clearing the way to enhance another's quality of life.

In addition to demolition, the federal government made urban investments through public housing development and the 1960s War on Poverty. But these programs did little to address patterns of racial inequality already built into the urban landscape. Public housing sites (e.g., Chicago's Robert Taylor Homes,

Pruitt-Igoe in St. Louis[7]) were segregated spaces (Hirsch 2009) situated in correspondingly racialized neighborhoods, preserving cities' already segregated landscapes. Only a portion of urban renewal displacees found homes in public housing. Many displaced people moved in with family or friends in nearby housing, increasing overcrowding – the opposite of the urban renewal program's stated goals (Sugrue 1996). Meanwhile, the "War on Poverty" channeled new resources into cities through programs like the Community Action Program division of the Office of Economic Opportunity. These programs emphasized the formal inclusion of low-income residents into decisions over the use of federal monies. Initiatives coming out of Community Action Programs focused on assimilating low-income people into the mainstream economy (e.g., reforming community cultures, individual behavioral reform, reorganization of social services). In Black urban communities, this approach failed to recognize or address the conditions of structural racism keeping people in poverty (O'Connor 2009).

Concurrently, the civil rights movement had recognized housing equity as a priority, campaigning to end discrimination in housing. The fight for fair housing drew national attention with the Chicago Freedom Movement, beginning in 1965, co-led by Martin Luther King Jr., which sought to dismantle segregation and discrimination in Chicago housing, schools, and labor markets (Finley et al. 2016). Locally, though, civil rights activists had been organizing for fair housing much earlier, specifically against urban renewal projects in their communities. Throughout the 1950s and 1960s, Black residents organized by the civil rights movement protested the destruction urban renewal brought to cities like Detroit, Oakland, and Atlanta (see, for example, Self 2005; Thomas 2013). In 1968, in response to Martin Luther King Jr.'s assassination and the subsequent uprisings in cities across the country, Congress passed the Fair Housing Act, which made racial discrimination in housing illegal but lacked meaningful enforcement mechanisms.[8] Despite these federal policy changes, research shows that housing discrimination (e.g., steering practices in which people are shown different properties based on race) remains widespread (Christensen and Timmins 2018).

1965–1980: reacting to urban crisis

By the late 1960s, it became apparent that urban planners' efforts were failing to renew cities economically and were unduly punishing the racialized poor. As living and working conditions for the African American poor stagnated or worsened, uprisings against racial oppression unfolded (called the "long hot summers" by white-narrating institutions). Black communities in many urban centers, building upon networks of civil rights organizing and influenced by emergent Black Power ideology, worked to radically reconfigure city-making processes. The Black Power Movement sought to achieve Black self-determination, emphasizing "the right to create our own terms through which to define ourselves, and our relationship to society, and to have these terms recognized" (Carmichael 1969). This idea of self-determination was central to a diverse and wide-ranging array of social change organizations: the Black Panther Party, the Student Nonviolent Coordinating Committee, the League of Revolutionary Black Workers, and the Third World Women's Alliance, among others. The Black Power movement inspired local organizing in cities throughout the United States around economics, education, welfare, prison/policing, electoral politics (strategically), and Pan-African solidarity. Influenced by the ideas of Frantz Fanon, Black Power nationalists and its leading edge, the Black Panther Party, conceptualized the Black community as an "internal colony" that was continuously struggling in the city for self-determination (Fanon 1966). The movement's central tenet: that Black people should control developments in their neighborhoods (Johnson 2007). This ran contrary to the top-down organization of planning at the time, which had imposed destruction and displacement on Black neighborhoods through the urban renewal regime. The effort to expand self-determination took many forms, including building power in urban government institutions through electoral politics (building top-down executive power), organizing independent schools and community organizations, and agitating for local reforms to education, social services, tenants' rights, and police reform (Collier-Thomas and Franklin 2001; Johnson 2007).

Planning as a field began to reckon with its evident failings for people of color and was forced to adjust (Thomas and Ritzdorf 1997). A rhetoric of race and class inclusiveness arose that permeated a litany of new planning and policy tools to remake cities: the community development block grant, urban homesteading, and urban development action grants. Planners began casting themselves as technicians sensitive to the needs of diverse people and communities, particularly the poor. Thus, these programs incorporated new mechanisms of community participation. For example, Community Development Block Grants, a federal grant provided to cities, required significant resident participation in deciding project priorities. Yet little changed. Planners continued to act as privileged influencers of city-making processes due to their formal training and technocratic knowledge. This expert knowledge about what was best for cities was based on assumptions that perpetuated race and class hierarchies: urban aesthetics and livability were best staked out through white middle-class sensibilities, poverty supposedly resulted from aberrant cultural practices, and economic growth always had to be given highest planning priority (VanBuskirk 1972; Dommel and Hall 1982).

This turn to the rhetoric of equity and community participation enabled urban policymakers to moderate Black urban political movements that sought to disrupt the economic and social systems privileging white America. This was exemplified in the formalization of community development in the 1970s. The federal government and philanthropic foundations, in response to urban uprisings and demands for radical social change in Black communities (Perry 1972), invested money in grassroots organizations – often highly politically active organizations that also provided for people's basic welfare – which would become "community development corporations." Federal urban policymakers strategically designed this investment to funnel resources to marginalized neighborhoods in a manner that would prevent future urban rebellions. As the federal government invested in these institutions, pressure was placed on these organizations to moderate their agendas (especially on economic matters), encouraging community development organizations to take guidance from powerful politicians and corporate leaders (Berdt 1977). Federal support for community development can be seen as an effort to respond to the demands of the Black poor while simultaneously curtailing the more radical aspects of movements for self-determination and community control. Investment in community development did enable more participatory, neighborhood-based planning, but it intentionally fell far short of the Black Power Movement's vision for radically reorganized, community-controlled neighborhoods.

1980–2000: hollowing out of the Black metropolis

1980s and 1990s planning in the United States was profoundly shaped by the rise of neoliberalism. This ideology emphasized the need to reduce government involvement in city affairs, trim especially the politics of redistribution (notably welfare provision), and allow the private market to allocate resources based on merit. City dwellers were assumed to all have unfettered opportunities to advance their own economic standing, provided they made wise decisions. Collectively, a new imagined city crystallized, one that relied on market-based solutions, reduced state welfare obligations, and cultivation of individual responsibility (Hackworth 2006).

In this context, planning efforts were built around servicing the abstract figure *homo economicus*, an individual whose actions were guided by economic rationality in a market economy (Rutland 2018). As urban planners increasingly viewed people primarily as economic actors (rather than encompassing the wholeness of the human experience), they defined good planning practice around a white normative standard. *Homo economicus* was assumed to have the economic and social opportunities of the white population. Yet, as industry relocated out of central cities, African American urban populations disproportionately faced unemployment, often making ends meet through a combination of welfare assistance and participation in the informal economy (Anderson 1987). The white normative standard held Black residents' economic strategies as undesirable, and at times criminally damaging to city improvement (Rutland 2018). At the same time, urban planning actively denied responsibility for the racialized effects of their plans, drawing on notions of "color-blind" operations, which posited a reality of "race-neutral" planning efforts (Mele 2013).

As neoliberalism guided urban policy, city problems were often attributed to "the behavioral problems of the people who live there" (Goetz 1996, 540). As such, public policy focused on modifying and policing people of color's behaviors. For instance, neoliberal policymakers argued that welfare programs bred "dependence" on the state and hampered economic productivity. Black women, in particular, were labeled with the perversely resonant notion "welfare queens" who unfairly took advantage of public resources, falling well short of white, neo-liberal norms of self-reliance (Hancock 2004).[9] Such labeling rhetoric fueled the supplanting of federal welfare by workfare that tied assistance to waged labor and its supposed therapeutic effects (Peck 2001). Government programs that failed to modify behaviors but merely offered survival resources (now termed "handouts") were subject to funding reductions and privatization. Thus, in this period, public housing programs were retooled to become market-based, contributing to a sig-nificant reduction in affordable housing units across urban America (Goetz 2012). Reductions in welfare resources hit Black communities especially hard. After dec-ades of disinvestment and job discrimination, Black neighborhoods most often in need of welfare resources had the least capacity to support locally-funded programs to mitigate federal policy changes (Moffitt 1998).[10]

At the same time, as welfare resources were withdrawn from cities with large Black populations, federal monies were channeled into cities through new polic-ing mechanisms as part of the "war on crime" and the later "war on drugs." Criminal behavior was identified as a primary factor preventing cities from flourishing (Goetz 1996). This tough-on-crime era disproportionately labeled the Black spaces and bodies of African-American communities as criminal and furthered Black Americans' mass incarceration (Covington 1995; Subramanian, Riley, and Mai 2018). The axis of anti-Black urban powers formed by the "war on drugs/war on crime" campaign made policing integral to urban planning efforts in housing and community development. For example, in the 1990s, "weed-and-seed" revitalization strategies integrated community development ("seeding" community resources) and policing ("weeding" out criminal influ-ences). While federal monies were withdrawn for programs that met survival needs, housing initiatives and community organizations could receive financial support to join police in surveilling people's behaviors, especially criminalized drug use (e.g., drug-testing programs, police-supported neighborhood watch programs, etc.) (Goetz 1996). In this way, planning became more integrated with urban policies that sought to discipline and punish those whose behavior failed to meet the white normative standard of *homo economicus*.

Between the criminalization of racialized poor and withdrawal of welfare resources, neoliberal urban policy and planning practices represent a regres-sion in the pursuit of racial equity. While earlier initiatives, like the War on Poverty and the community development movement, aspired to eliminate the causes of race-class inequality, neoliberal policy blamed the racialized poor for urban problems. Urban policy went from "fighting poverty to fighting the poor" (Goetz 1996, 548).

2010–present: Black right to the city matters

In recent years, planners have centered urban cores as sites for investment and gentrification. Urban planners and policymakers have encouraged gentri-fication through a litany of initiatives: providing tax incentives, designating historic areas, scripting tax increment financing, facilitating "greening aesthetics," and implementing upscale rezoning (D. Wilson 2018). This embrace of gentrification has had racialized effects. As neighborhoods – often those near the anchor institutions constructed during urban renewal (universities, hospitals, etc.) (see Taylor Jr and Cole 2001) and increasingly those in climate resilient areas ("climate gentrification") – become redeveloped for more affluent consumption, costs of living rise, and people of color are frequently displaced. In many cities, gentrification dislocates African Americans from neighborhoods that have been centers of Black cultural life for decades. For example, between 2000 and 2013 in Washington DC, Black residents were displaced from 60% of gentrifying neighborhoods (Richardson, Mitchell, and Franco 2019). Even when displacement does not occur, remaining people live more precarious lives due to increases in rent, food prices, and local property taxes (D. Wilson 2018).

In readying neighborhoods for gentrification, planners and policymakers continue to integrate police into planning practices. Alliances between planners, policymakers, and police around "broken windows" initiatives vividly illustrate this. The theory: preventing minor acts of social disorder (e.g., loitering, drinking, spitting) prevents more serious degradation of neighborhoods (J. Q. Wilson and Kelling 1982). Guided by this theory, urban planners believe that by controlling visualities of disorder, neighborhoods could be made more suitable to real estate investment and gentrification (Harcourt and Ludwig 2006). The repercussions are not surprising: Black people going about their lives are often interpreted by police and policymakers as visual signs of disorder (see Conquergood 1991) in gentrifying areas. In the process, broken windows theory contributes to police harassment of African Americans in and near gentrifying urban neighborhoods (Laniyonu 2018).

Yet resistance has arisen to such over-policing with people organizing locally, nationally, and internationally under the banner of Black Lives Matter. As this movement advocates for defunding policing (reallocating local police budgets to other community programs) and abolition of the prison industrial complex, it advances very different visions for urban futures. These visions now prompt a new reckoning within urban planning. The field is once again confronted with the damage many planners have done to Black populations. Reflecting on the police killing of George Floyd in 2020, Julian Agyeman (2020) has identified not just policing but planning as a primary shaper of structural racism in America:

> The legacy of structural racism in Minneapolis was laid bare to the world at the intersection of Chicago Avenue and East 38th Street, the location where George Floyd's neck was pinned to the ground by a police officer's knee. But it is also imprinted in streets, parks and neighborhoods across the city – the result of urban planning that has used segregation as a tool of white supremacy.

A fundamental question is on the table: can planners contribute to creating a racially just city, and if so, how?

Some planners have called for the development of abolitionist planning in which planners work to dismantle institutions that produce "systematic racialized political and economic violence against people of color (Abbot et al. 2018)." For example, the abolitionist planner would work to counter the commodification of housing and urban landscapes, processes that disproportionately afflict Black lives. The abolitionist approach rejects color-blind neoliberal planning and seeks to redirect planning practice toward truly community-led, community-controlled initiatives.[11] But, similar calls have been made before, with planners forced to reconcile the consequences of their actions. And while those moments of reflection led to modest changes to equity in planning, planning practice largely continues to serve the interests of the powerful. As such, others argue that the goal of abolition is fundamentally incompatible with professionalized planning practice. For example, Dozier (2018) has argued that professional planning is too deeply connected to dominant power systems and focused on moderate reforms to contribute to the abolition movement. Abolition planning may not be work that can be done in the institutions where planners usually work – institutions that are more likely to offer up rhetorical support for equity than meaningful transformation (Dozier 2018).[12] Like earlier movements for Black liberation, the Black Lives Matter Movement offers significantly different visions for the future of cities and challenges planners committed to racial justice to radically reimagine their work.

Conclusion

This chapter has revealed the contradiction that lies at the heart of the story planning often tells about itself: that planning improves all city dwellers' lives. Throughout history, urban planners in the United States have repeatedly positioned Black communities and their residents as impediments to city progress. In the process, planning has helped produce and reproduce the structures of racism in American society that continue to award social and material benefits to white populations. Urban planning and policy have, in an ongoing way, restricted racialized people's residential options, displaced them from their homes and communities, dismantled their wealth, and aided in producing the inequality that marks contemporary urban America. Today, white families have a significantly higher median household income ($65,000 vs. $39,000) and household wealth ($171,000 vs. $17,000) than Black families (V. Wilson 2018). Black Americans are almost six times as likely to be incarcerated as white Americans (Carson 2020). Increased exposure to environmental toxins in urban environments has contributed to poorer health outcomes for Black Americans, including in the COVID-19 pandemic (Noppert 2020).

Urban planning persists as a technocratic instrument that unequally generates and distributes benefits to populations. While many urban planners hope that their work will contribute to making more racially just cities, racial capitalism to date has imposed immense constrictions on the planning apparatus. It asks planners to sort out who will go where, under what conditions, and for whose benefit under the constraint that cities are unwavering sites for capital accumulation. This ongoing drive for capital accumulation powerfully limits urban planners' actions and prevents them from redressing the racial inequalities structured into cities. Still, we recognize that city futures have not been determined; demands for urban planning and policy reform are being made throughout the country. A diversity of resistance movements – in housing, welfare provision, fair employment, immigrant rights – continue to challenge the dominant planning paradigm to deliver resources to historically neglected communities. However, structures of power, including white supremacy, are not easily supplanted. Now is the long-overdue moment for planning institutions to respond to demands from within and outside the profession to concretely and significantly transform the racial politics of the discipline (e.g., enacting the numerous racial equity reforms proposed by faculty of color over the past ten years in planning education programs highlighted by Acey et al. 2020).[13] This will be no small feat. Achieving racially just cities will require a radical reimagining of what urban planning should and could be.

Alternative planning history and theory timeline

Period	Planning history and theory canon	Alternative timeline: people of color
Late 1800s– 1900	Birth of planning	*Planning in the Jim Crow era* Sanitary and housing reforms neglect drivers of racial inequality
1900–1945	Formalization of planning	*Urban planning builds the ghetto* Production of the segregated city through zoning and redlining
1945–1965	Growth of planning	*Planning through demolition* Destruction of Black communities through urban renewal and highway construction
1965–1980	Midlife crisis of planning	*Reacting to urban crisis* Black Power organizing for community control, planning's pivot to community development
1980–2000	Maturation of planning	*Hollowing out of the Black metropolis* Dismantling of the welfare state, rise of mass incarceration as urban policy
2010–present	New planning crisis	*Black right to the city matters* Planning and racial justice in the gentrifying city

Case study: Racial space-making, environmental justice, and Black right to the city

This case study presents the case of Altgeld Gardens, a public housing development on the far Southside of Chicago. Altgeld Gardens' history exemplifies the intergenerational legacies of housing segregation and the subsequent organizing to realize the demand for a Black right to the city that continues to this day around environmental justice. In the World War II era, Chicago planners sought to expand the city's housing supply to accommodate the wartime industrial workforce and, later, returning soldiers. This was a racially bifurcated project: whites primarily channeled toward homeownership with the aid of GI Bill mortgage benefits (a benefit not extended to Black soldiers),[14] and Black residents funneled into new public housing projects in separate parts of the city. Thus, as Chicago expanded its housing, it maintained its racially segregated form. Altgeld Gardens was one such segregated, public housing development. It was constructed in 1944–1945 to house Black World War II industry workers,[15] as well as returning GIs and their families, in southeast Chicago near defense factories. The self-contained "garden city" design of the project (including a public library, schools, a daycare, and urban gardening green space) was promoted as progressive urban design close to new industrial employment opportunities (Preservation Chicago 2020). Ironically, this allegedly progressive expansion of Chicago's racial segregation would later produce the lead plaintive in the historic housing desegregation Supreme Court "Gautreaux Case," Altgeld Gardens resident Dorothy Gautreaux (Hal Baron 1991). The 1976 Supreme Court case would find that the Chicago Housing Authority's public housing siting practices violated the civil rights act which resulted in the Gautreaux Project, the largest housing desegregation program in the U.S. history.

Within decades of its construction, Altgeld Garden's proximity to industry would prove harmful to residents. Over time, "the Gardens" became surrounded by 50 landfills and 382 industrial facilities. The area surrounding Altgeld Gardens has been characterized as a "toxic doughnut," containing high concentrations of hazardous industrial pollutants and toxic waste. Toxicology studies report dangerous levels of mercury, lead, DDT, PCBs, PAHs, heavy metals, and xylene (People for Community Recovery 2020). In the 1970s, Altgeld residents, led by Hazel M. Johnson, later dubbed the "Mother of the Environmental Justice Movement," began investigating connections between pollutants in their living environment and the high rates of cancer on Chicago's far Southside. In 1979, Johnson helped establish the neighborhood organization, People for Community Recovery (PCR), which identified pollutants affecting the housing development. In response, over the past 40 years, PCR has conducted periodic surveys of residents' health, organized residents to demand change from corporate polluters, held government agencies accountable for enforcing environmental standards, and educated Altgeld Gardens residents about environmental health risks (Chicago Public Library 2014). Johnson and PCR joined activists across the United States to bring the environmental concerns of communities of color to the attention of the predominately white environmental movement and the federal government. Johnson was part of a coalition of activists who influenced President Clinton to sign the 1994 Environmental Justice Executive Order, which required federal agencies to identify and address adverse environmental conditions in minority and low-income communities (Chicago Public Library 2014). People for Community Recovery continue to lead the fight for environmental justice in Altgeld Gardens today. The Altgeld Gardens case makes clear the longue durée of racist planning: segregation in one generation has led to environmentally produced, racialized health inequalities in subsequent generations. However, urban movements motivated by the idea of Black rights to the city have also proved persistent. The work of grassroots environmental organizations like People for Community Recovery have not only made inroads to secure the right to the city for their neighbors, but also helped to develop the principle of environmental justice itself, an idea which continues to guide visions for equitable cities.

Notes

1 In this chapter, we use the words Black and African American interchangeably to refer to members of the African diaspora living in the United States. We note that not all Black people living in the United States identify as African American. For example, immigrants from Africa, the Caribbean, and Latin America may identify themselves based on their nation of origin rather than adopting the term African American.

2 Hal Baron and Henry Taylor have each argued that the outcomes of racial inequalities produced by planning are posited as "neutral and value-free." That is the nature of institutional, structural, or color-blind racism. The neutral, value-free assumptions, methods and practices taught in much of planning pedagogy are anything but.

3 There are disturbing echoes of the 1918 "Spanish flu" pandemic that met Black urban migrants from the South in the 2020 COVID-19 pandemic. The racial disparities of this folded-history of public health crises reveal that Black city-dwellers did not so much succumb from disparate infection rates as from the racial discrimination of city public health agencies and hospitals. The coronavirus has folded the history Black healthcare disparities onto the same urban geography of spatial segregation from a century earlier.

4 In 1968, Black Chicagoans organized the Contract Buyers League to oppose the real estate industry's contract selling practices (Coates 2014).

5 The GI Bill established a range of benefits (e.g., low-interest home loans) to assist returning World War II veterans, thus enabling many white families to buy suburban homes. The Federal Highway Act of 1956 launched the creation of the interstate highway system. Suburban communities were built along newly constructed highways since the infrastructure offered speedy commutes to employers in city centers.

6 Funding for demolition began in 1949, but the language of urban renewal was not invoked until the Housing Act of 1954.

7 Pruitt-Igoe was a famously short-lived public housing development, opening in 1954 and demolished by 1976. Residents have cited a lack of maintenance investment as the cause of the project's failure (see *The Pruitt-Igoe Myth: An Urban History*). There has been a chronic lack of maintenance funding for public housing developments in the United States, contributing to poor living conditions in many of these communities. In recent years, Public Housing Operating and Capital Funds (the federal government fund for basic maintenance of public housing facilities) has only been fully funded twice since 2002.

8 One of the most incisive criticisms of the racism inherent in planning's urban renewal strategy was lodged in the famous 1968 Kerner Commission Report on Civil Disorder released one month before the assassination of Dr. King.

9 The power of misogynoir masked the reality that whites formed the largest population receiving welfare.

10 The simultaneity of decades of disinvestment in community development and over-investment in policing to suppress the societal effects of severely under-resourced communities climaxed in a new cycle of repression and resistance. The cycle of police brutalities and extra-judicial killings of Black residents stemming from this community disinvestment/policing over-investment dialectic has produced a new stage of Black self-determination in 2020 with the Black Lives Matter Movement's call to "defund the police."

11 While the abolition planning approach shares some similarities to Paul Davidoff's advocacy planning (i.e., an understanding that planning is political and should be used to promote equity), they are not synonymous. The concept of advocacy planning focused on changing the role of the planner from a rational actor producing technical documents to a political actor advancing the cause of marginalized groups in the planning process. Abolition planning has a specific focus on dismantling institutions that produce racialized harm and replacing them with new, non-violent systems.

12 For a recent example of this dynamic in planning education see Acey et al. (2020).

13 These recommendations have included incorporating issues of diversity into core curriculum requirements (not just elective courses), addressing discrimination and bias directed toward students of color in planning, and diversifying the faculty and student bodies of planning programs.

14 Public housing developments were also created for white Chicagoans in Chicago's segregated public housing system.

15 World War II era industrial jobs were opened to African Americans by President Roosevelt's signing Executive Order 8802 creating the Fair Employment Practices Committee to desegregate war industries under threat of the March on Washington Movement organized by Black civil and labor rights leaders A. Phillip Randolph and Bayard Rustin.

References

Abbot, Thomas, Roxana Aslan, Riley O'Brien, and Nathan Serafin. 2018. "Embrace Abolitionist Planning to Fight Trumpism." Progressive City. April 6, 2018. https://www.progressivecity.net/single-post/2018/04/06/EMBRACE-ABOLITIONIST-PLANNING-TO-FIGHT-TRUMPISM.

Acey, Charisma, Lisa Bates, Andrew Greenlee, Michael C. Lens, Willow Lung-Amam, Andrea Roberts, Sheryl-Ann Simpson, et al. 2020. "Black Faculty Response to the ACSP Statement." Association of Collegiate Schools of Planning. June 18, 2020. https://www.acsp.org/news/511293/ACSP-Statement-Following-the-Killing-of-George-Floyd.htm#Black%20Faculty%20Response.

Agyeman, Julian. 2020. "Urban Planning as a Tool of White Supremacy – the Other Lesson from Minneapolis." The Conversation. July 27, 2020. https://theconversation.com/urban-planning-as-a-tool-of-white-supremacy-the-other-lesson-from-minneapolis-142249.

Anderson, Bernard. 1987. "The Changing Workplace and Unions: Implications for Black Workers." In *The Changing Economy and Unions: An Analysis and Program for the Black-Labor Alliance*, 3–35 Washington, DC: A. Philip Randolph Institute.

Baron, Hal. 1991. "What Is Gautreaux? The Name of a Gallant Woman." In *What Is Gautreaux?*, edited by Business and Professional People for the Public Interest, Chicago, 1–2.

Baron, Harold. 1968. "Planning in Black and White." In *The Racial Aspects of Planning: An Urban League Critique of the Chicago Comprehensive Plan*, edited by Harold Baron. Chicago: Chicago Urban League Research Report.

Berdt, H. 1977. *New Rulers in the Ghetto: The Community Development Corporation and Urban Poverty.* Westport, CT: Greenwood Press.

Bullard, Robert Doyle, Glenn Steve Johnson, and Angel O. Torres. 2004. *Highway Robbery: Transportation Racism & New Routes to Equity.* Cambridge, Massachusetts: South End Press.

Carmichael, Stokely. 1969. "Stokely Carmichael Explains Black Power to a White Audience in Whitewater, Wisconsin." In *The Rhetoric of Black Power.* New York: Harper & Row.

Carson, E. Ann. 2020. "Prisoners in 2018." Bulletin NCJ 253516. U.S. Department of Justice. Office of Justice Programs. Bureau of Justice Statistics. https://www.bjs.gov/content/pub/pdf/p18.pdf.

Cha-Jua, Sundiata Keita. 2000. *America's First Black Town: Brooklyn, Illinois, 1830–1915.* Urbana: University of Illinois Press.

Chicago Public Library. 2014. "People for Community Recovery Archives." 2014. https://www.chipublib.org/fa-people-for-community-recovery-archives/.

Christensen, Peter, and Christopher Timmins. 2018. *Sorting or Steering: Experimental Evidence on the Economic Effects of Housing Discrimination.* Cambridge, Massachusetts: National Bureau of Economic Research.

Coates, Ta-Nehisi. 2014. "The Case for Reparations." *The Atlantic* 313 (5): 54–71.

Collier-Thomas, Bettye, and V. P. Franklin. 2001. *Sisters in the Struggle: African American Women in the Civil Rights-Black Power Movement.* New York University Press, New York: NYU Press.

Connolly, Nathan DB. 2014. *A World More Concrete: Real Estate and the Remaking of Jim Crow South Florida.* Vol. 114. Chicago: University of Chicago Press.

Conquergood, Lorne Dwight. 1991. *Life in Big Red: Struggles and Accommodations in a Chicago Polyethnic Tenement.* Vol. 91, 2. Center for Urban Affairs and Policy Research, Evanston: Northwestern University.

Corn, Marti, Tracy Xavia Karner, Thad Sitton, Tacey A. Rosolowski, and Wanda Horton-Woodworth. 2016. *The Ground on Which I Stand: Tamina, a Freedmen's Town.* College Station, United States: Texas A&M University Press. http://ebookcentral. proquest.com/lib/ilstu/detail.action?docID=5843520.

Covington, Jeanette. 1995. "Racial Classification in Criminology: The Reproduction of Racialized Crime." *Sociological Forum* 10 (4): 547–68.

Digital Scholarship Lab (2016), "Renewing Inequality." *In American Panorama*, edted by Robert K. Nelson and Edward L. Ayers. https://dsl.richmond.edu/panorama/renewal/#view=0/0/1&viz=cartogram.

Dommel, Paul R., and John Stuart Hall. 1982. *Decentralizing Urban Policy: Case Studies in Community Development.* Washington DC: Brookings Institution Press.

Dozier, Deshonay. 2018. "A Response to Abolitionist Planning: There Is No Room for 'Planners' in the Movement for Abolition." Planners Network. August 9, 2018. https://www.plannersnetwork.org/2018/08/response-to-abolitionist-planning/.

Drake, Saint C., and Horace R. Cayton. 1970. *Black Metropolis: A Study of Negro Life in a Northern City.* Chicago: University of Chicago Press.

Du Bois, William Edward Burghardt. 1899. *The Philadelphia Negro: A Social Study*, 14. Philadelphia: University of Pennsylvania Press

Fanon, Frantz. 1966. *The Wretched of the Earth (1st Evergreen Ed.).* New York: Grove Weidenfeld.

Farrar, Hayward. 1998. *The Baltimore Afro-American, 1892–1950*, 185. Westport Connecticut: Greenwood Publishing Group.

Finley, Mary Lou, Bernard Jr. Lafayette, James R. Jr Ralph, and Pam Smith, eds. 2016. *The Chicago Freedom Movement: Martin Luther King Jr. and Civil Rights Activism in the North*. Lexington: University Press of Kentucky.

Fullilove, Mindy Thompson. 2016. *Root Shock: How Tearing up City Neighborhoods Hurts America, and What We Can Do about It*. New Village Press.

George, S., A. Hendley, Jack Macnamara, J. Perez, and A. Vaca-Loyola. 2019. "The Plunder of Black Wealth in Chicago: New Findings on the Lasting Toll of Predatory Housing Contracts." *Durham, NC: Duke University, The Samuel Du Bois Cook Center on Social Equity. Retrieved June* 1: 2019.

Goetz, Edward G. 1996. "The US War on Drugs as Urban Policy." *International Journal of Urban and Regional Research* 20 (3): 539–49. https://doi.org/10.1111/j.1468-2427.1996. tb00332.x.

———. 2012. "The Transformation of Public Housing Policy, 1985–2011." *Journal of the American Planning Association* 78 (4): 452–63. https://doi.org/10.1080/01944363.2012. 737983.

Goetz, Edward G., Rashad A. Williams, and Anthony Damiano. 2020. "Whiteness and Urban Planning." *Journal of the American Planning Association* 86 (2): 142–56. https:// doi.org/10.1080/01944363.2019.1693907.

Hackworth, Jason. 2006. *The Neoliberal City: Governance, Ideology, and Development in American Urbanism*, 1st ed. Ithaca: Cornell University Press.

Hancock, Ange-Marie. 2004. *The Politics of Disgust: The Public Identity of the Welfare Queen*. New York: NYU Press.

Harcourt, Bernard, and Jens Ludwig. 2006. "Broken Windows: New Evidence from New York City and a Five-City Social Experiment." *University of Chicago Law Review* 73 (1). https://chicagounbound.uchicago.edu/uclrev/vol73/iss1/14.

Hawthorne, C. (2019). Black matters are spatial matters: Black geographies for the twenty-first century. *Geography Compass*, 13(11), e12468. https://doi.org/10.1111/ gec3.12468

Hirsch, Arnold R. 2009. *Making the Second Ghetto: Race and Housing in Chicago 1940–1960*. Chicago: University of Chicago Press.

Jackson, Kenneth T. 1987. *Crabgrass Frontier: The Suburbanization of the United States*. New York: Oxford University Press.

Johnson, Cedric. 2007. *Revolutionaries to Race Leaders: Black Power and the Making of African American Politics*. Minneapolis: University of Minnesota Press.

Laniyonu, Ayobami. 2018. "Coffee Shops and Street Stops: Policing Practices in Gentrifying Neighborhoods." *Urban Affairs Review* 54 (5): 898–930. https://doi. org/10.1177/1078087416689728.

Levy, John M. 2016. *Contemporary Urban Planning*. New York: Taylor & Francis.

"Mapping Inequality: Redlining in New Deal America." 2016. American Panorama. Digital Scholarship Lab, University of Virginia. https://dsl.richmond.edu/panorama/ redlining/#loc=5/39.1/-94.58&text=intro.

McKittrick, K., & Woods, C. A. (2007). *Black geographies and the politics of place*. South End Press. Cambridge, Massachusetts.

Mele, Christopher. 2013. "Neoliberalism, Race and the Redefining of Urban Redevelopment." *International Journal of Urban and Regional Research* 37 (2): 598–612.

Moffitt, Robert A. 1998. *Welfare, the Family, and Reproductive Behavior: Research Perspectives*. National Research Council Committee on Population, Washington DC: National Academies Press.

Nightingale, Carl H. 2012. *Segregation: A Global History of Divided Cities.* Chicago: University of Chicago Press.

Noppert, Grace A. 2020. "COVID-19 Is Hitting Black and Poor Communities the Hardest." JSTOR Daily. April 14, 2020. https://daily.jstor.org/covid-10-hitting-black-poor-communities-hardest/.

O'Connor, Alice. 2009. *Poverty Knowledge: Social Science, Social Policy, and the Poor in Twentieth-Century US History.* Princeton: Princeton University Press.

Osofsky, Gilbert. 1971. *Harlem: The Making of a Ghetto: Negro New York, 1890–1930,* 3. New York: HarperCollins College Division.

Peck, Jamie. 2001. *Workfare States.* New York: The Guilford Press.

People for Community Recovery. 2020. "History." People for Community Recovery. 2020. http://www.peopleforcommunityrecovery.org/history.html.

Perry, Stewart E. 1972. "Black Institutions, Black Separatism, and Ghetto Economic Development (Includes Comment and Author's Reply)." *Human Organization* 31 (3): 271–79.

Peterson, Jon A. 2003. *The Birth of City Planning in the United States, 1840–1917.* Baltimore: JHU Press.

Preservation Chicago. 2020. "Altgeld Gardens." 2020. https://preservationchicago.org/chicago07/altgeld-gardens-blocks-11-12-and-13/.

Rabin, Yale. 1989. "Expulsive Zoning." In *Zoning and the American Dream: Promises Still to Keep.* Vol. 101, 105. New York: The American Planning Association, Planners Press.

Richardson, Jason, Bruce Mitchell, and Juan Franco. 2019. *Shifting Neighborhoods: Gentrification and Cultural Displacement in American Cities.* Washington DC: National Community Reinvestment Coalition.

Ritzdorf, Marsha. 1997. "Locked out of Paradise: Contemporary Exclusionary Zoning, the Supreme Court, and African Americans, 1970 to the Present." In *Urban Planning and the African American Community: In the Shadows,* edited by June Manning Thomas and Marsha Ritzdorf. Thousand Oaks, CA: Sage.

Roithmayr, Daria. 2010. "Racial Cartels." *Mich. J. Race & L.* 16: 45.

Rothstein, Richard. 2017. *The Color of Law: A Forgotten History of How Our Government Segregated America.* New York: Liveright Publishing.

Rutland, Ted. 2018. *Displacing Blackness: Planning, Power, and Race in Twentieth-Century Halifax.* Toronto: University of Toronto Press.

Schlabach, Elizabeth. 2019. "The Influenza Epidemic and Jim Crow Public Health Policies and Practices in Chicago, 1917–1921." *The Journal of African American History* 104 (1): 31–58. https://doi.org/10.1086/701105.

Self, Robert O. 2005. *American Babylon: Race and the Struggle for Postwar Oakland.* Vol. 34. Princeton: Princeton University Press.

Shabazz, Rashad. 2015. *Spatializing Blackness: Architectures of Confinement and Black Masculinity in Chicago.* Urbana: University of Illinois Press.

Silver, Christopher. 1997. "The Racial Origins of Zoning in American Cities." In *Urban Planning and the African American Community: In the Shadows,* edited by June Manning Thomas and Marsha Ritzdorf. Thousand Oaks, CA: Sage.

Stein, Samuel. 2019. *Capital City: Gentrification and the Real Estate State.* Verso Books.

Subramanian, Ram, Christine Riley, and Chris Mai. 2018. *Divided Justice: Trends in Black and White Jail Incarceration, 1990–2013.* New York: Vera Institute of Justice.

Sugrue, Thomas J. 1996. *The Origins of the Urban Crisis Race and Inequality in Postwar Detroit.* Princeton: Princeton University Press.

Talen, Emily. 2014. "Housing Demolition during Urban Renewal." *City & Community* 13 (3): 233–53.

Taylor Jr, Henry Louis, and Sam Cole. 2001. "Structural Racism and Efforts to Radically Reconstruct the Inner-City Built Environment." In *ACSP Fannie Mae Foundation Award Best Action Research Paper 2001 43 Rd Annual Conference, Association of Collegiate Schools of Planning*. Retrieved June 1, 2005.

Teaford, Jon C. 1990. *The Rough Road to Renaissance: Urban Revitalization in America, 1940–1985.* Baltimore: Johns Hopkins University Press.

Thomas, June Manning. 2013. *Redevelopment and Race : Planning a Finer City in Postwar Detroit/.* Paperback ed. Detroit, MI: Wayne State University Press.

Thomas, June Manning, and Marsha Ritzdorf, eds. 1997. *Urban Planning and the African American Community: In the Shadows.* Thousand Oaks: Sage.

Trotter, Joe, Earl Lewis, and Tera Hunter. 2004. *The African American Urban Experience: Perspectives from the Colonial Period to the Present.* New York: Springer.

Trotter, Joe William, ed. 1991. *The Great Migration in Historical Perspective: New Dimensions of Race, Class, and Gender.* Vol. 669. Bloomington: Indiana University Press.

United States Bureau of Labor. 1897. "Condition of the Negro in Various Cities." *Bulletin of the United States Bureau of Labor II* (10): 257–369.

VanBuskirk, Paul. 1972. *The Resurrection of an American City.* Morristown, NJ: Schenkman.

Von Hoffman, Alexander. 1998. *The Origins of American Housing Reform*, 2. Joint Center for Housing Studies, Cambridge, Massachusetts: Harvard University.

Weiss, Marc Allan. 1980. "The Origins and Legacy of Urban Renewal." in Pierre Clavel, John Forester, and William W. Goldsmith, eds., *Urban and Regional Planning in an Age of Austerity.* Pergamon Press. New York.

Wilson, David. 2018. *Chicago's Redevelopment Machine and Blues Clubs.* Cham, Switzerland: Springer.

Wilson, James Q., and George L. Kelling. 1982. "Broken Windows." *Atlantic Monthly* 249 (3): 29–38.

Wilson, Valerie. 2018. "Racial Inequalities in Wages, Income, and Wealth Show That MLK's Work Remains Unfinished." *Economic Policy Institute* (blog). 2018. https://www.epi.org/publication/racial-inequalities-in-wages-income-and-wealth-show-that-mlks-work-remains-unfinished/.

Suggestions for further study

- Gordon, Colin. 2008. Mapping Decline: St. Louis and the American City. Interactive Web Map. http://mappingdecline.lib.uiowa.edu/.
- Hunter, Marcus Anthony. 2013. *Black Citymakers: How the Philadelphia Negro Changed Urban America.* Oxford University Press.
- Thomas, J. M. 2013. *Redevelopment and Race: Planning a Finer City in Postwar Detroit.* Wayne State University Press.

9

MIGRANTS

Anti-Mexicanism and the elusive American dream

Álvaro Huerta, Enrique M. Buelna, and Gabriel Buelna

Introduction

To better understand the history of the United States one must examine the urban history of all racialized and marginalized groups therein. While the struggles of American Blacks are much better documented and publicly known, knowledge of racist planning policies (historical and contemporary) against individuals of Mexican origin (including U.S. citizens and recent migrants) is desperately lacking.

The historical ties between Mexico and the United States, especially the Southwest, distinguish Mexican immigrants and Mexican Americans (Chicanas/os) from other ethnic communities. Unlike Europeans, who chose to cross the Atlantic Ocean in order to settle in North America, Mexicans have always been here – given their indigenous roots. Hence, Mexican migration toward *el norte* cannot be considered as "illegal," given that historically, the Southwest of the United States was in fact Mexican land.

Despite Mexicans' tremendous economic, cultural, artistic and social contributions to U.S. cities, they continue to be demonized as "criminals," "drug dealers," and "rapists" (Reilly 2016). Even those people of Mexican origin that have acquired U.S. citizenship are treated as "foreigners in their own land" (Takaki 2008, 165). Gómez-Quiñones (2019) posits:

> U.S. anti-Mexicanism is a race premised set of historical and contemporary ascriptions, convictions and discriminatory practices inflicted on persons of Mexican descent, longstanding and pervasive in the United States ... Anti-Mexicanism is a form of nativism practiced by colonialists and their inheritors ...

DOI: 10.4324/9781003157588-10

In his classic book, *A Different Mirror: A History of Multicultural America*, Ronald Takaki documents how white Americans gradually migrated into Mexican territory starting in the 1820s, specifically in what is now Texas. At the time, the Mexican government allowed these settlers to enter under the assumption that they would adopt Mexican customs, learn Spanish and intermarry with locals. Coexistence occurred without much conflict until 1836 when white American settlers revolted and secured independence for Texas through violence. Once Texas joined the United States, it did not take the American government long to expand its territory into the Southwest by waging war against Mexico from 1846 to 1848 (Acuña 2004; Huerta 2019). This bloody land grab was based on the false notion of Manifest Destiny: "a religious doctrine with roots in Puritan ideas, which continue to influence U.S. thought to this day" (Acuña 2004, 52). Once the United States and Mexico signed the Treaty of Guadalupe Hidalgo in 1848, Mexico lost half of its territory (Meyer, Sherman, and Deeds 1999; Takaki 2008).

Mexican migration to the United States must be examined against this background.[1] In their analysis of the spatial structure and the ethno-racial inequality in Los Angeles over time, Ong and R. Gonzalez (2019, 190) summarize some of the issues that are unpacked in this chapter:

> … Hispanics have been a part of Los Angeles well before California was seized by the US government. Through much of that period, they have been economically marginalized. As with Asians, the long history of marginalization of Hispanic individuals and neighborhoods in Los Angeles contribute to the inequality. For instance, the racial stereotyping of youth during Zoot Suit Riots post-World War II and stigmatizing of life in the 'barrio.' The marginalization stems back even further to the Great Depression when protectionist policies such as the repatriation of Mexicans was a popular response to the deepening economic problems in Los Angeles. The confounding issue of citizenship, legality, and the right to remain has not disappeared. In today's charged political climate, Latinos are marginalized as undesirable immigrants and undocumented aliens.

Late 1800s–1900: U.S. invasion of Mexico and rise of anti-Mexican sentiment

The study of urban planning ultimately consists of the relationship between people, land, and the built environment. In the case of individuals of Mexican origin in the United States, it starts with the U.S. annexation of half of Mexico's territory in the mid-19th century. As with Native American treaties, the United States broke the Treaty of Guadalupe Hidalgo: Mexicans who had previously settled in *el norte* – what is now known as Southwest of the United States – did not receive equal protections to white Americans. Instead they were treated as a servant class from the outset.

Nonetheless, the presence of people of Mexican origin had a positive (albeit often uncredited) impact on American cities, towns, and villages during the mid- to late-19th century. Mexicans contributed greatly to many areas of American society and economy. For example, during the second half of the 19th century, Mexican immigrants and their offspring (Mexican Americans) constituted the key labor force in agriculture, railroad construction, and mining (Takaki 2008).

However, instead of being rewarded for their contributions with opportunities for upward mobility, they experienced much racism in the workplace. For example, Mexican workers in Anglo-owned ranches in Texas, "found themselves in a caste system – a racially stratified occupational hierarchy" (Takaki 2008). While Mexicans occupied the lower-paid laborer ranks, white Americans worked as supervisors or managers. This workforce stratification by ethnicity, along with a segregated educational system, limited the upward mobility of Mexican Americans (and immigrants) in America from the start of the invasion and annexation.

1900–1945: dispossession, removal, and precarious status

By the turn of the 20th century, individuals of Mexican origin had been largely dispossessed of the millions of acres of land they had once owned. The process of dispossession included a combination of (a) judicial malfeasance, such as denial of due process, rejection of grant rights, and reduction of land titles; (b) nefarious legislative measures such as tax laws, squatters' rights, gerrymandering, and poll taxes; and (c) high legal costs, sealed off resources, and competition with land companies (Acuña 2004; Takaki 2008). Where these actions failed to dislodge Mexican Americans from their property holdings, death threats, property destruction, murder (lynch mobs), and wholesale massacres were deployed, often under the watch of local and state authorities. For Mexican Americans and Mexican immigrants, extrajudicial floggings, beatings, mutilations, and killings became part and parcel of life in the United States at the cusp of the 20th century (Acuña 2004; Carrigan and Webb 2013; Vargas 2017). Some ended up working as farm hands on the very lands they once owned.

Even those Mexicans who managed to retain their titles saw their lives upturned. The shift from mercantile capitalism to industrial capitalism brought on tremendous changes that did not benefit the majority of Mexican Americans who had long resided in the Southwest. Many were forced to abandon farming and rural areas and join the expanding industrial order in cities, if only for short stints. Most became itinerant workers, moving to cities where work was available, even if that meant traveling long distances to neighboring states.

Recent migrants did not fare much better. As commercial agriculture, mining, and railroad construction expanded, and immigration from Asia and Europe was closed off or declined, industrial capital looked to Mexico to meet its demand for cheap and docile labor. As a result, tens of thousands of Mexican migrants entered the United States (Foley 1997). They hoped for better economic opportunities but harsh conditions awaited them. Once again, they were cast into the lower rungs of the labor and racial order, being viewed as expendable outsiders, unworthy of citizenship. Any attempt at self-determination was viewed by Anglo Americans as an act of defiance and a potential threat to the ethnic status quo. Organized violence was often employed as a means to weaken labor organizing, cut short demands for wage increases, thwart strikes, and reestablish or expand control of the geographic landscape (Acuña 2007; Vargas 2017).

However, for Mexicans migration was preferable to life back home. Under the authoritarian regime of Porfirio Díaz (1876–1880 and 1884–1911), Mexico underwent a modernization program, which involved development policies that benefited foreign investors and wealthy landowners. This program expelled thousands of poor rural residents from their lands, forcing a mass of impoverished, landless peasantry to migrate to larger urban centers within Mexico and the United States (Acuña 2007).

The vast majority of migrant workers in the U.S. faced racism, discrimination, harsh working conditions, and a dual wage system. Some settled in company towns where rents were high whereas others were in constant movement which made life very hard for families (Acuña 2007; Gordon 1999). Indeed, Anglo antipathy toward Mexican Americans ("the mongrel breed"; Vargas 2017, 163) was so deep that it delayed statehood for Southwestern territories (New Mexico and Arizona) until 1912.

Despite their precarious status, Mexican Americans and immigrant workers still managed to organize, often joining African Americans and other ethnic groups in making successful demands of their employers. With the outbreak of the revolution in Mexico in 1910, violence along the U.S. border escalated to new heights. Between 1915 and 1919, an estimated 5,000 men, women, and children were killed in Texas's lower Rio Grande Valley in what can only be described as an ethnic war (Carrigan and Webb 2013; Vargas 2017). Prompted initially by a revolt of Tejanos – Texans of Mexican descent – demanding the liberation of Mexico's former territories, the war came to be known as *Hora de Sangre* (Hour of Blood) (Carrigan and Webb 2013; Martinez 2018; Montejano 1987). The resulting trauma affected Mexican Americans for many years.

Anti-Mexicanism must be viewed in the context of growing apprehension over the shifting ethnic and racial makeup of the American population starting in the early 1900s. The white portions of the population feared that their supremacy might be challenged by Mexicans, Asians, and other so-called "inferior" ethnicities, in addition to African Americans. Eugenics, the pseudo-science that purported to use the principles of genetics to improve the human race, fueled the notion that Mexicans were a "problem" population. In a 1930 report submitted to Congress, economist Roy L. Garis of Vanderbilt University characterized the presence of Mexican Americans and Mexican immigrants as an "invasion" that, if left unresolved, would lead the nation "perilously near the brink of mongrelism." Indeed, Garis worried that Mexicans' "indifference to racial intermixture," coupled with their pride and sensitivity to racial prejudice, would lead to ongoing ethnic conflict (Acuña 2004; Stepan 1991; U.S. Congress 1930). The practice of "redlining," entire neighborhoods as high-risk areas making it very costly to secure home loans, dating back to the 1930s, targeted Mexican barrios, as well as African American communities (Domonoske 2016).

While Mexican Americans were increasingly identified as "undesirables," economic exigencies prevailed against any move to have them removed. In Southern California, white Americans tolerated the presence of a large Mexican American population through a "racial logic" lens: Mexican workers were perceived as being both expendable and necessary (González 1994; Molina 2006; Vargas 1999). The attitude of businesses such as Simons Brick Company in Los Angeles was typical: "More Mexicans meant more workers. More workers meant more bricks. More bricks meant more money" (Buelna 2019).

But in 1929, at the onset of the Great Depression, Mexican Americans and other immigrant communities were once again targeted as the nation searched for scapegoats. The clamor to remove "foreigners" who took away scarce jobs and other resources from "locals" affected every family of Mexican descent across America. Authorities made little distinction between American or foreign-born people. In one community after another, Mexican-looking people were picked up at schools, work sites, hospitals, stores, parks, and street corners. Packed into buses, squad cars, or taxis, they were transported to processing centers or straight to the border. No allowances were made to arrange care for children left behind, to collect wages, or to sell property. By the end of the 1930s, between half a million and one million Mexicans were repatriated. According to some estimates, the numbers could have been as high as two million, of which two-thirds were U.S.-born children. These uprooted Mexican American children found themselves in a country that they hardly knew. Mexico, in the midst of its own economic crisis, did not welcome them with open arms (Acuña 2004; Balderrama and Rodríguez 2006; Gutiérrez 1995; Rodriguez 2007; Romo 1983).

As the Depression progressed, a younger generation of Mexican Americans stepped in to fill the void left behind by the removals, emboldened by a new spirit of civic activism. They joined the expanding working-class movement taking shape across the United States and took leadership roles in the new federation of labor, the Congress of Industrial Organizations, as well as New Deal agencies and radical grassroots organizations. As labor organizers and leaders, Mexican Americans and other Latino groups helped lead workers' struggles in cities across the United States. These experiences, in turn, gave them the confidence to address broader community concerns, such as poor housing, healthcare, and schools (Buelna 2019; Denning 1997; García 1991; Sánchez 1993). During this period, segregated schools for Mexican American and Mexican immigrant children were very common in the U.S. cities and rural communities – although teachers were not necessarily of Mexican origin. These schools, which had inferior buildings and teaching standards, set children up to fail and then conditioned them to accept failure (Acuña 2004).

Despite the outpouring of civic activism, at the start of WWII Mexicans remained suspect, segregated into urban barrios, and constantly under surveillance. While an estimated 500,000 Mexican Americans joined the armed forces during the war, their service and sacrifice was not sufficient to dislodge long-held racial and ethnic prejudices. In wartime Los Angeles, for example – and along the West Coast in general – race became a litmus test for loyalty and true citizenship. In 1942, hysteria over internal enemies exploded with accusations that Mexican Americans were part of a secretive "fifth column," working to undermine national security. The local media framed Mexican American youth as "rat packs," gangs of free-roaming criminals sowing urban chaos and undermining civil society. Control over these youth became imperative, often necessitating large dragnets that terrorized entire Mexican and other minority communities (Buelna 2019; Escobar 1999; Gutiérrez 1995; Mazón 1984). In June 1943, a ten-day riot erupted in Los Angeles, largely attributed to Anglo servicemen, who were unwilling to share the city's geography with Mexicans, a group they deemed inferior and foreign. Similar altercations with U.S. service personnel occurred across California. The Los Angeles Police Department, not unlike law enforcement agencies across the nation, strove to maintain the city's racial and ethnic status quo. In effect, the message for Mexican Americans, even in the midst of a national emergency, was clear: racial and ethnic divisions were there to stay (Buelna 2019).

However, the wartime environment created some breaches in the system, which young *pachucos* – a particular Mexican American urban youth subculture of the era – learned to navigate and exploit. The wartime economy had allowed *pachucos* greater opportunities to work and earn money. They could move more freely about the city, challenging racial barriers that had once kept them out of certain indoor and outdoor spaces. They frequented new businesses, assembled freely in new places, and consumed American culture in new ways. The trademark *pachuco* style was the zoot suit – an outfit also worn by working-class youth of other ethnicities. This was *pachucos'* way of breaking free from the pressure to conform to both Mexican family norms and Anglo American society. It also offered a means to express the dual nature of their Mexican and American experience. Eventually, the zoot suit was racialized as a style of clothing associated exclusively with Mexican Americans. As *pachucos* became more visible and increasingly confident in their identity, Anglo Americans framed them as examples of "Mexican degeneracy" and, ultimately, as internal threats (Escobar 1999; Mazón 1984; Sánchez 1993).

1945–1965: urban renewal, racial segregation, and displacement

In the United States, the post-World War II era is generally considered as a period of economic growth and development. Well recovered from the Depression and the war, the government engaged in a massive urban renewal program, which involved slum clearance and redevelopment and the expansion of urban and interurban freeways (Acuña 2004; Badger 2016; Cebul 2020). These planning approaches led to suburbanization and a major white flight from the cities to the suburbs. Brent Cebul (2020) describes urban renewal as "… ceding to the interests of commercial lobbyists and mayors. The [Housing Act of 1954] loosened housing mandates and so unleashed a goldmine of federally financed business-oriented clearance and development."

Throughout the 1950s, urban renewal caused the mass displacement of Latinos (and other racial minorities) from their barrios in urban cores (Badger 2016; Cebul 2020; Estrada 2005). By the early 1960s, more than 600,000 people had been uprooted nationally, two-thirds of whom were minority members. Displacement hardly helped improve their standard of living. The experience of Chicanas/os in Los Angeles was typical. Huante (2021, 65) describes the segregationist forces at play in this major center of Mexican-American culture:

> Keeping ethnic Mexicans spatially contained and placing them at the lower rungs of the racial hierarchy has represented an ongoing project for Los Angeles civic leaders since the Treaty of Guadalupe Hidalgo of 1848 … Los Angeles' rise as a prominent American city was largely attributed to cultural and civic labor's investment in "whitewashing" the city's Mexican past and articulating an Anglo futurity … As Los Angeles' racial landscape took shape, Anglo boosters, civic leaders, tourists, and residents participated in daily practices based on an imagined map of the city that considered particular spaces as foreign and unsettled, necessitating the fabrication of social distance between themselves and these spaces … As a result, Anglo newcomers were directed to the west side of the L.A. civic center, while ethnic newcomers to the city remained confined to areas such as Boyle Heights–that is, east of the L.A. River …

When freeway plans were proposed, the sections of the city where Chicanas/os lived were considered expendable. The government used the power of eminent domain to remove these groups so that financial interests could reap large profits. Acuña (2004, 295) describes the case of a local community in his classic book, *Occupied America: A History of Chicanos*:

> Encouraged by this renewal process, Mexican communities were kept in a state of flux throughout the 1950s as they became targets of developers. In October 1958 the city removed Mexican homeowners from Chávez Ravine, near the center of Los Angeles giving more than 300 acres of private land to Walter O'Malley, owner of the Dodger baseball team. The Dodgers deal angered many people; residents of Chávez Ravine resisted physically. In 1959 the county sheriff's department forcibly removed the Aréchiga family. Councilman Ed Roybal condemned the action: 'The eviction is

the kind of thing you might expect in Nazi Germany or during the Spanish Inquisition.' Supporters of the Aréchigas protested to the City Council. Victoria Augustian, a witness, pointed a finger at city council member Rosalind Wyman, who with Mayor Norris Poulson supported the giveaway. Poulson was a puppet of the Chandler family, who owned the *Los Angeles Times*, which backed the handover.

Many life-long Mexican American residents of Los Angeles retain a collective memory of this incident (Acuña 2004; Shatkin 2018). While baseball is a favorite American pastime, for many Mexican Americans whose family members were displaced from Chávez Ravine, the idea of cheering for the local team, the Los Angeles Dodges, triggers a painful historical memory.

1965–1980: rise of the Brown Power movement

In the turbulent 1960s, many Mexican Americans claimed the term *Chicanas/os* and demanded civil and human rights. Young and rebellious, the Chicanas/os rejected the assimilationist views of their parents' generation. Rather, they sought self-determination. In addition to fighting against the disproportionate drafting of Chicanos in the Vietnam war, activists focused on key issues impacting urban communities. This included protesting against poor and still segregated schools, where Chicanas/os lacked opportunities to pursue higher education and achieve upward mobility. A major demonstration, attended by 20,000 people, took place in Los Angeles on August 29, 1970, which led to the police killing of the acclaimed Chicano journalist Ruben Salazar (Huerta 2019).

Grassroot organizations of that era included the Young Chicano for Community Action (YCCA), which eventually transformed into the Brown Berets, modeled after the Black Panther Party. Female activists, such as Gloria Arellanes, played a major role in these organizations (Montes 2020). Chicanas helped open and operate La Pirañya, a coffeehouse in East Los Angeles, which for years served as a hub for socializing, hosting speeches by community leaders such as Rodolfo "Corky" Gonzales, sharing news about politics, and organizing political activities. Los Angeles Chicanas also produced their own newspaper, *La Causa*, and operated El Barrio Free Clinic, a free medical facility for the community.

In addition to protesting in cities, Mexican Americans and immigrants also protested in rural areas, particularly in America's agricultural fields. Co-led by Cesar Chavez and Dolores Huerta, the United Farm Workers (UFW) was founded in 1962 in response to the inhumane working conditions of farmworkers in California and other states, such as Arizona, Texas, Florida, and Washington. The UFW organized and joined picket lines and marches, signed petitions, supported labor laws, lobbied elected officials, distributed educational flyers, produced documentaries, penned songs, performed plays, and held teach-ins. The charismatic Chavez – who appeared on the cover of *Time* magazine in 1969 – engaged in numerous and lengthy hunger strikes to draw attention to the cause of Mexican American farmers. Many UFW activists were beaten up and a few were killed. In 1970, the UFW won a major battle, when powerful grape growers in California agreed to sign the first contract with the union (Acuña 2004; Gómez-Quiñones 1990; Huerta 2019). At the occasion, Dolores Huerta famously proclaimed, "¡*Si, Se Puede!*" (Yes, we can!). The activism of Chicanas/os continued into the 1970s not only in the fields, but also in cities and education campuses at all levels.

1980–2000: the struggle against anti-Mexicanism continues

The period between 1980 and 2000 saw very stark differences in how the United States handled immigration policy at both the state and federal levels. While most of the 1980s were marked by the conservative policies of Republican President Ronald Reagan, this was a positive decade in terms of immigration reform. Reagan's attitude toward migrants was beneficial to Mexican and Latino communities. He supported a general amnesty for close to 2.7 million undocumented immigrants (Nyce and Bodenner 2016).

With the amnesty, Mexicans in America could finally enjoy a sense of peace and place, and no longer had to fear deportation. Families that were once worried about building wealth, acquiring property, attending school, or visiting families in Mexico and farther afield in Latin America, could now live with some level of tranquility (Badger 2014). Moreover, millions of Mexican and Latino immigrants became eligible to vote beginning in 1993. Ultimately, this led to significant increases in Mexican American and Latino political power in predominately urban states with large population centers (Pachon and DeSipio 1992).

While on the domestic front Reagan provided legal status to undocumented migrants, internationally he supported wars in El Salvador and Nicaragua which prompted millions of Central Americans to migrate to the United States. The chaos, strife, and distress caused by those wars has endured in Central America into the present (Shull 2021). Meanwhile, the flow of Salvadorian and Nicaraguan migration, along with the regular flow of Mexican migration, acted as a trigger to the ideological backlash of the following decade.

The 1990s saw the adoption of draconian state law initiatives targeting migrants. An argument could be made that the election of Donald Trump to the American Presidency in 2016 could not have occurred without the ideological foundation and the anti-immigrant rhetoric of the 1990s. Practically, the 1990s were a return to the old anti-Mexicanism of the late 1800s and early to mid-20th century (Gómez-Quiñones 2017), with legislation threatening deportations and limiting services based on Mexican heritage.

Among the first clearly anti-Mexican pieces of legislation were California Propositions' 187 and 227 – passed in 1994 and 1998, respectively (Bali 2001; Silva 2019). Proposition 187 banned services to undocumented migrants, including education and healthcare for children and families. The initiative found favor with voters (Arellano 2019), but was later found to be unconstitutional. By contrast, Proposition 227 focused on eliminating bilingual education as a teaching pedagogy in public schools. Chicanas/os and Latinas/Latinos saw this as a targeted move to reinforce Anglo cultural supremacy and undermine Mexicans' "stubbornness" in sustaining the Spanish language and Mexican/Latino heritage. Both Propositions were accompanied by blatant anti-Mexican campaigns, which framed migrants as freeloaders rather than as key contributors to the urban and rural economies of the United States.

The volume *In the Barrios: Latinos and the Underclass Debate* paints a vivid picture of Latino life and struggles in American cities at the end of the 20th century. The editors, Joan More and Raquel Pinderhughes (1993:xvi–xvii) note that:

> American minorities have been incorporated into the general social fabric in a variety of ways. Just as Chicago's black ghettos reflect a history of slavery, Jim Crow legislation, and struggles for civil and economic rights, so the nation's Latino barrios reflect a history of conquest, immigration, and a struggle to maintain cultural identity.

While the 1990s were a generally dark political period for Mexican American and Latino communities in the United States, elected political representation skyrocketed during this time period too – owing to the newly naturalized citizens. In California, by the early 2000s, Latino politicians constituted one fifth or more of both the state assembly and senate (Buelna, G. 2003, 98). This newfound political clout had a tangible effect on the physical environment of cities. For example, it was reflected in the approval of new mass transit systems and new school construction bonds in Los Angeles and throughout California.

2000–present: changing demographics and enduring discrimination of Mexican Americans

In the 21st century, Mexican Americans (and other Latino groups) have become a major political force by virtue of their sheer numbers. The largest racial-ethnic group in the United States, in 2000 Latinos already accounted for 12.5% of the total population with individuals of Mexican origin representing the largest subgroup. Two decades later, their numbers had reached nearly 61 million (or 18% of the total population).

Significant portions of Mexicans and other Latinos have assimilated into Anglo society, joined the middle and upper rungs of the socioeconomic ladder, and even acquired the status of "honorary whites" (Huante 2021; Tran and Valdez 2017). Note the similarity with Asian Americans: they too have been "upgraded" from an underclass to a "model minority" of hardworking, smart, rule-abiding, and self-reliant people (Zhang 2010). These stereotypes are hardly flattering, as they "other" migrants while concealing many issues, including racism, poverty, poor housing conditions, labor abuse, and psychological pressures to succeed in life without expecting assistance from the public largesse. Worse, they serve to pitch minority groups against one another.

In reality, poverty and segregation persist for millions. As South et al. (2005:580) point out,

> ... Latino immigrants who are embedded in networks of coethnics would be more likely to remain in, or move to, neighborhoods that are inhabited primarily by Latinos. For these immigrants, ethnic communities are likely to provide sources of material and emotional support, which in turn serve to retain immigrants in these communities.

However, working-class barrios are not chosen by Latinos simply for the "material and emotional support" that they provide. Poor brown migrants are often barred from white affluent neighborhoods through "invisible" financial, linguistic, and cultural barriers.

In spite of, or perhaps because of, the upward socioeconomic mobility of some Latinos, discrimination and inflammatory rhetoric has not ceased. The tumultuous period of Republican President Donald J. Trump (2017–2021) posed an existential threat not only to new Mexican arrivals, but also to citizens and permanent residents ("green card" holders) of Mexican origin. In 2015, during his presidential announcement speech, Trump stated:

> ... When Mexico sends its people, they're not sending their best. ... They're sending people that have lots of problems, and they're bringing those problems with us. They're bringing drugs. They're bringing crime. They're rapists ...
>
> (quoted in Huerta 2019; Reilly 2016).

A cornerstone of Trump's presidential campaign was the construction of a wall along the U.S.-Mexico border to keep potential migrants out. Once in office, he supported inhumane immigration policies including mass deportations of brown immigrants, separation of migrant families along the entire U.S.-Mexico border, and even detention of migrant children. Trump's defeat in the 2020 presidential elections was met with a sigh of relief by most Mexican Americans and Mexican immigrants; these communities played a key role in his electoral defeat.

However, "Trumpism" was probably a symptom rather than a cause of xenophobia and anti-Mexicanism. The reality is that during the administration of Democratic President Barack Obama (2009–2017) more than 2.7 million immigrants were deported as well (Huerta 2019). This shows that both major political parties in the United States strongly believe in enforcement-only measures when it comes to immigration policy. As of this writing, there are about 11 million undocumented immigrants in urban and rural areas who lack a pathway citizenship; the majority are of Mexican origin.

There is hope that the new Democratic administration of President Joseph Biden, which came into power in 2021, will support Latino communities in U.S. cities through progressive housing, transport, education, immigration, and healthcare policies/programs. Much optimism stems from the fact that the current Vice-President, Kamala Harris, is the first woman, the first person of color, and the first daughter of migrants to have achieved such a high office in the United States.

However, in addition to racism and xenophobia, the forces of neoliberal capitalism – on the rise in the United States since the 1980s – present a barrier to the well-being and advancement of Mexican Americans and Mexican migrants, especially poorer ones, regardless of the political party in power. For example, in the San Francisco Bay Area, the rise of Silicon Valley as a global technology hub attracted an influx of highly paid white-collar employees from all over the country, leading to the hyper-gentrification of the city of San Francisco and the displacement of impoverished, disempowered tenants from traditionally Mexican neighborhoods such as the Mission District.

Under these circumstances, it is imperative that people of Mexican origin, Latinos all as well as other minority groups, organize from the ground up to create novel, more positive narratives around migration and demand pro-migrant policies from urban leaders. Too often, public policy change takes place only when the citizenry demands it.

Conclusion

By focusing on the case of anti-Mexicanism in U.S. cities, this chapter expands the narrative of racism and discrimination beyond the white-Black paradigm. While it is imperative for all minority and oppressed groups to unite and support each other's causes, planners must also have a comprehensive understanding of all racialized groups and their particular histories. This is critical because, at various times, the planning profession has actively contributed to the marginalization, segregation and subjugation of Mexican Americans and Mexican immigrants, among others. At this stage in U.S. history, as individuals of Mexican origin continue to experience system racism at all levels, planners need to take an active role in re-imagining more just and inclusive urban communities. Anti-Mexicanism can only be considered defeated in the United States when Latino minorities not only cease to be targets of racist rhetoric, but also when they come to enjoy all the benefits that developed American cities can offer. This includes access to affordable housing, quality public schools, comprehensive healthcare, good jobs, and opportunities for upward mobility. In other words, racial justice and equity in the United States will only occur when Mexican Americans and Mexican immigrants (as other racialized groups) have equal access to the increasingly elusive American Dream.

Alternative planning history and theory timeline

Period	Planning history and theory canon	Alternative timeline: migrants
Late 1800s–1900	Birth of planning	*U.S. Invasion of Mexico and rise of anti-Mexican sentiment* Colonization, land theft, broken Treaty, occupational hierarchy
1900–1945	Formalization of planning	*Dispossession, removal, and precarious status* Labor subjugation, residential segregation
1945–1965	Growth of planning	*Urban renewal, racial segregation, and displacement* Highway projects cutting across barrios
1965–1980	Midlife crisis of planning	*Rise of the Brown Power movement* Chicanas/os seeking self-determination, grassroot organizations, protests in cities and rural areas
1980–2000	Maturation of planning	*Continuing struggle against anti-Mexicanism* Immigration reform (1980s), draconian state law initiatives (1990s), barrios as repositories of Latino cultural identity
2000–present	New planning crisis	*Changing demographics* Browning of America, assimilation, Latinos as "honorary whites," enduring discrimination of poor Mexican Americans, barrio gentrification

Case study: the informal economy of Los Angeles

In Los Angeles, many Mexican immigrants, who are shut out from the formal economy, survive by pursuing economic activities in the informal sector. In 2004, around 679,000 people were part of the informal economy in Los Angeles County. Their jobs tended to be temporary or occasional, with earnings often not reported (in part or in full) to the taxation office. The wages were rather slim relative to the high cost of living in Los Angeles: only $12,000 per year on average based on a 2005 estimate by a local organization, Economic Roundtable. However, given a mass of Mexican migrants, the total amount earned informally amounted to $8.1 billion per year. Typically, workers and petty entrepreneurs were paid in cash (or sometimes personal checks) and they received no benefits such as health insurance, workers compensation, maternity leave, or pension funds. The goods and services provided by the informal economy are mentioned here because they are very relevant to urban planning. Migrants are involved in (a) food preparation and serving, both in restaurants and outdoor food trucks;

(b) street vending, which serves poorer suburbs where access to "formal" stores is lacking; (c) small construction projects, such as house renovations, as day laborers; and (d) domestic work such as cleaning houses, looking after children and the elderly, or tending gardens. In other words, they have taken on the type of work that is critical to the well-functioning of the city but that white American denizens tend to reject as "dirty." However, Mexican immigrants regard this as honest work which provides them with an income and some dignity in a competitive and costly urban setting. Not only does their hard-earned "informal" money allow them to support themselves and their families in Los Angeles without asking for charity but remittances also support the families left behind in Mexico and sustain the economy of Mexican cities and towns.

Note

1 A note on the terms employed in this chapter, in addition to Mexican or Mexican-American:
  Hispanic: a term used to refer more specifically to Spanish-speaking people in the United States, especially of Latin American descent. Brazilians are Latinos but not Hispanic.
  Chicano/a: a term that refers to Americans (men or women) of Mexican origin or descent.

References

Acuña, Rodolfo F. 2007. *Corridors of Migrations: The Odyssey of Mexican Laborers, 1600–1933.* Tucson: University of Arizona Press.

Acuña, Rodolfo F. 2004. *Occupied America: A History of Chicanos.* Fifth Edition. New York: Pearson Longman.

Arellano, Gustavo. 2019. "Prop. 187 Forced a Generation to Put Fear Aside and Fight. It Transformed California, and Me." *Los Angeles Times*, October 29, 2019.

Badger, Emily. 2014. "What Happened to the Millions of Immigrants Granted Legal Status Under Ronald Reagan?" *The Washington Post*, November 26.

Badger, Emily. 2016. "Why Trump's Use of the Words 'Urban Renewal' Is Scary for Cities." New York Times, December 7.

Balderrama, Francisco E. and Raymond Rodríguez. 2006. *Decade of Betrayal: Mexican Repatriation in the 1930s.* Albuquerque: University of New Mexico Press.

Bali, Valentina A. 2001. "Sink or Swim: What Happened to California's Bilingual Students after Proposition 227?" *State Politics & Policy Quarterly* 1(3): 295–317.

Budiman, Abby, Tamir, Christine, Lauren Mora and Noe-Bustamente, Luis. 2018. "Statistical Portrait of the Foreign-Born Population in the United States." *Pew Research Center, Hispanic Trends, Report.*

Buelna, Enrique M. 2019. *Chicano Communists and the Struggle for Social Justice.* Tucson: The University of Arizona Press.

Carrigan, William D. and Clive Webb. 2013. *Forgotten Dead: Mob Violence Against Mexicans in the United States, 1848–1928.* New York: Oxford University Press.

Cebul, Brent. 2020. "Tearing Down Black America." Black Agenda Report, July 29, 2020.

D'Antonio, Michael. 2017. "Trump's Move to End DACA Has Roots in America's Long, Shameful History of Eugenics." *Los Angeles Times*, September 14.

Dear, Michael. 2013. *Why Walls Won't Work: Repairing the US-Mexico Divide*. Oxford, England: Oxford University Press.

Denning, Michael. 1997. *The Cultural Front: The Laboring of American Culture in the Twentieth Century*. London: Verso.

Domonoske, Camila. 2016. "Interactive Redlining Map Zooms in on America's History of Discrimination." *NPR*, October 19, 2016.

Escobar, Edward J. 1999. Race, Police, and the Making of a Political Identity: Mexican Americans and the Los Angeles Police Department, 1900–1945. Berkeley: University of California Press.

Estrada, Gilbert. 2005. "If You Build It, They Will Move: The Los Angeles Freeway System and the Displacement of Mexican East Los Angeles, 1944–1972." *Southern California Quarterly* 87(3): 287–315

Economic Roundtable. 2005. *Hopeful Workers, Marginal Jobs: LA's Off-the-Books Labor Force*. Report, Los Angeles.

Foley, Neil. 1997. *The White Scourge: Mexicans, Blacks, and Poor Whites in Texas Cotton Culture*. Berkeley: University of California Press.

García, Mario T. 1991. *Mexican Americans: Leadership, Ideology, and Identity, 1930–1960*. New Haven: Yale University Press.

Gómez-Quiñones, Juan. 2017. "On the Chicana/o Moratorium of 1970: A Legal Historical Inquiry Into the Events of the Day and the Death of Ruben Salazar, a Journalist Killed." *HuffPost*, August 22, 2017.

Gómez-Quiñones, Juan. 1990. *Chicano Politics: Reality and Promise 1940–1990*. Albuquerque: University of New Mexico.

Gómez-Quiñones, Juan. 1994a. *Mexican American Labor: 1790–1990*. Albuquerque: University of New Mexico.

Gómez-Quiñones, Juan. 1994b. *Roots of Chicano Politics, 1600–1940*. Albuquerque: University of New Mexico.

Gómez-Quiñones, Juan. 2019. "La Realidad: The Realities of Anti-Mexicanism – A Paradigm." In *Defending Latina/o Immigrant Communities: The Xenophobic Era of Trump and Beyond*, edited by Alvaro Huerta, pp. 9–13. Lanham: Roman and Littlefield.

Gonzales, Manuel G. 2009. *Mexicanos: A History of Mexicans in the United States*. Second Edition. Bloomington: Indiana University Press.

González, Gilbert G. 1994. *Labor and Community: Mexican Citrus Worker Villages in a Southern California County, 1900–1950*. Urbana: University of Illinois Press.

Gordon, Linda. 1999. *The Great Arizona Orphan Abduction*. Cambridge: Harvard University Press.

Grossman, Roger. 2017. "The 1954 Deportation of Mexican Migrants and the 'Wetback Airlift' in Chicago." *Chicago Tribune*, March 4.

Gutiérrez, David G. 1995. *Walls and Mirrors: Mexican Americans, Mexican Immigrants, and the Politics of Ethnicity*. Berkeley: University of California Press.

Guzman, Betsy. 2001. Census 2000 Brief: The Hispanic Population. May 1.

Huante, Alfredo. 2021. "A Lighter Shade of Brown? Racial Formation and Gentrification in Latino Los Angeles." *Social Problems* 68(1): 63–79.

Huerta, Alvaro and Gabriel Buelna. 2021. "Explaining Anti-Mexicanism-Racism Toward Mexican Americans and Latinos." *BuelnaNews*, February 24, 2021.

Huerta, Alvaro. 2019. *Defending Latina/o Immigrant Communities: The Xenophobic Era of Trump and Beyond*. Lanham: Roman and Littlefield.

Huerta, Alvaro. 2020. "When Mexican Americans Transformed into Chicanos and Chicanas: Remembering East L.A.'s Chicano Moratorium 50 Years Ago." *L.A. Taco*, August 27.

Huntington, Samuel P. 2004. "The Hispanic Threat." *Foreign Policy* 141: 30–45.

Martinez, Jack. 2015. "Republican Democrats Debate Immigration as Trump Praises Eisenhower's 'Operation Wetback.'" *Newsweek*, November 10.

Martinez, Monica Muñoz. 2018. *The Injustice Never Leaves You: Anti-Mexican Violence in Texas.* Cambridge: Harvard University Press.

Mazón, Mauricio. 1984. *The Zoot-Suit Riots: The Psychology of Symbolic Annihilation.* Austin: University of Texas Press.

Menchaca, Martha. 1995. *The Mexican Outsiders: A Community History of Marginalization and Discrimination in California.* Austin: University of Texas Press.

Meyer, Michael C., William L. Sherman and Susan M. Deeds. 1999. *The Course of Mexican History.* Sixth Edition. Oxford, England: Oxford University Press.

Mimbs Nyce, Caroline and Bodenner, Chris. 2016. "Looking Back at Amnesty Under Reagan." *The Atlantic*, May 23.

Molina, Natalia. 2006. *Fit To Be Citizens? Public Health and Race in Los Angeles, 1879–1939.* Tucson: University of Arizona Press.

Montejano, David. 1987. *Anglos and Mexicans in the Making of Texas, 1836–1986.* Austin: University of Texas Press.

Montes, Carlos. 2020. Personal interview conducted by Alvaro Huerta, March 20, 2020.

Moore, Joan and Pinderhughes, Raquel, eds. 1993. *In the Barrios: Latinos and the Underclass Debate.* Thousand Oaks, CA: Russell Sage Foundation.

Noe-Bustamente, Mark Hugo Lopez and Jens Manuel Krogstad. 2020. "U.S. Hispanic Population Surpassed 60 Million in 2019, but Growth Has Slowed." *Pew Research Center*, July 7.

Ong, Paul M. and Silvia R. Gonzalez, 2019. *Uneven Urbanscape: Spatial Structures and Ethnoracial Inequality.* Cambridge, UK: Cambridge University Press.

Pachon, Harry and Louis DeSipio. 1992. "Latino Elected Officials in the 1990s." *PS: Political Science and Politics* 25 (2): 212–217.

Reilly, Katie. 2016. "Here Are all the Times Donald Trump Insulted Mexico." Time, August 31.

Rodriguez, Gregory. 2007. *Mongrels, Bastards, Orphans, and Vagabonds: Mexican Immigration and the Future of Race in America.* New York: Pantheon Books.

Romo, Ricardo. 1983. *East Los Angeles: History of a Barrio.* Austin: University of Texas Press.

Sánchez, George. 1993. *Becoming Mexican American: Ethnicity, Culture, and Identity in Chicano Los Angeles, 1900–1945.* New York: Oxford University Press.

Shatkin, Elina. 2018. "The Ugly, Violent Clearing of Chavez Ravine Before It Was Home to the Dodgers." *LAlist*, October 17.

Shull, Kristina. 2021. "Reagan's Cold War on Immigrants: Resistance and the Rise of a Detention Region, 1981–1985." *Journal of American Ethnic History* 40 (2): 5–51.

Silva, Andrea. 2019. "How California's Prop. 187 Is Still Shaping Immigration Policy— 25 Years after It Passed." *The Washington Post*, November 25, 2019.

South, S.J., Crowder, K. and Chavez, E. 2005. "Geographic Mobility and Spatial Assimilation among U. S. Latino Immigrants." *International Migration Review* 39: 577–607.

Stepan, Nancy Leys. 1991. *"The Hour of Eugenics": Race, Gender, and Nation in Latin America.* New York: Cornell University Press.

Takaki, Ronald. 2008. *A Different Mirror: A History of Multicultural America.* Revised Edition. New York: Back Bay Press.

Torres, Rodolfo D. and David R. Diaz, eds. 2021. *Latino Urbanism: The Politics of Planning, Policy, and Redevelopment.* New York: New York University Press.

Tran, Van C. and Nicol M. Valdez, 2017. "Second-Generation Decline or Advantage? Latino Assimilation in the Aftermath of the Great Recession." *International Migration Review* 51: 155–190.

United States. 1930. *Western hemisphere immigration Hearings before the Committee on Immigration and Naturalization, House of Representatives, Seventy-first Congress, second session, on the bills, H.R. 8523, H.R. 8530, H.R. 8702, to limit the immigration of aliens to the United States, and for other purposes. Printed for the use of the Committee on Immigration and Naturalization.* Washington: U.S. Govt. Print. Off.

Vargas, Zaragosa. 2017. *Crucible of Struggle: A History of Mexican Americans from Colonial Times to the Present Era.* New York: Oxford University Press.

Zhang, Qin. 2010. "Asian Americans Beyond the Model Minority Stereotype: The Nerdy and the Left Out." *Journal of International and Intercultural Communication* 3(1): 20–37.

Suggestions for further study

- Anzaldua, Gloria. 2012. *Borderlands/La Frontera*: The New Mestiza. Fourth Edition. 25th Anniversary. San Francisco: Aunt Lute Books.
- Bracero History Archive: http://braceroarchive.org.
- Gómez-Quiñones, Juan. "Thirteen Rhetorical Notes: A theoretical departure, *la noche triste*, a defeat of the Europeans by the Aztecs—Equality, justice, liberty, freedom, harmony and neither genocide, imprisonment or slavery." YouTube, September 17, 2020. https://www.youtube.com/watch?v=R2z80pKJWWg.
- Gómez-Quiñones, Juan. 2012. *Indigenous Quotient/Stalking Words: American Indian Heritage as Future.* San Antonio: Aztlan Libre Press.
- UCLA Chicano Studies Research Center: https://www.chicano.ucla.edu.

10

INDIGENOUS PEOPLE

Māori planning rights and wrongs in Aotearoa

Meg Parsons

Introduction

Aotearoa New Zealand (hereafter Aotearoa) is undeniably a Māori space in which hundreds of different iwi (tribes) and hapū (sub-tribes) continue to maintain their intergenerational connections to their ancestral whenua (lands), awa (rivers), and moana (seas) despite more than 150 years of colonial rule (Anderson, Binney, and Harris 2015). Thus, all planning practices that have taken place in Aotearoa since British colonization formally began in 1840 occurred on the rohe of iwi; rohe is the Māori word that describes the territorial lands, waters, and seas of a particular iwi or hapū.

While Māori interests, priorities, and knowledge (mātauranga Māori) feature more and more in debates about current planning practices, these remain in the periphery of mainstream planning activities (Livesey 2019). In Aotearoa, the discipline of planning continues to ignore how colonial foundations and practices are interwoven into the histories of Māori dispossession and their unjust relations with the settler colonial state (Matunga 2013). These histories and experiences reverberate across Aotearoa and continue to shape how planning is conceptualized and enacted in the nation.

To critically examine the histories and practices of urban planning in Aotearoa, one must situate knowledge and power at the heart of settler efforts to control and remake land. This chapter draws from scholarship from settler-colonial, feminist, decolonial, and Indigenous studies (Barry and Thompson-Fawcett 2020; Livesey 2019; Matunga 2013; Porter, Jackson, and Johnson 2017; Veracini 2010; Wolfe 2006).

In his 2006 work, Patrick Wolfe maintained that settler colonialism is "rooted in the elimination of the Indigenous peoples, polities and relationships from and with the land" (Wolfe 2006, 7–8). The settler–colonial project in Aotearoa was

DOI: 10.4324/9781003157588-11

founded on the dispossession of the Indigenous Māori people of their lands, waters, flora, and fauna, as well as other things (material and metaphysical) that were of cultural significance to them. Hegemonic discourses and racialized inferences continue to be evident in the lived realities of colonial spaces and planning regimes.

Colonial planning traditions, transplanted from Britain and adapted to fit the Antipodean settings, engaged in early practices of spatial planning that focused on the "taming" the "wilderness" (Parsons et al. 2019). These included activities such as surveying, mapping, selecting and acquiring land (through stealing, purchasing, confiscating), defining spaces for particular social groups and activities, clearing vegetation, as well as (re)naming places, imposing new regimes of land tenure and land use, and constructing towns, infrastructure, factories, and houses (Byrnes 2001). Settler-colonial planning practices, Matunga (2013, 9) argues, excluded and marginalized Indigenous planning approaches:

> the materiality (i.e. physical quality, presence, and structure) and memory (i.e. recall of existence, even existence) of Indigenous communities has generally been erased. In the cities, it was replaced with imperial monuments, colonial buildings, colonial cathedrals, colonial gardens, and colonial city patterns modelled on the 'old world' and the mother country [...]. The aim was to remove any material evidence/reminder and memory of Indigenous communities, their places, sites, resources and villages, and replace it with a new colonial order.

The dispossession of Māori in Aotearoa from their tribal homelands (what Māori refer to as their rohe) is what Oren Yiftachel calls "the dark side of planning" (Yiftachel 1998, 385). Dispossession, as Porter et al. (2017, 20) summarize, is the "forced removal of territory from a sovereign people." Although the pattern of dispossession varied widely across Aotearoa, colonial violence and discriminatory policies against Māori, and the denial of Māori rights to self-determination were experienced by all iwi and hapū (Parsons et al. 2019). The settler-colonial project in Aotearoa, paralleling those in Australia and North America, was ultimately about securing land (and other resources) for settlers and supplanting Indigenous society with settler society. However, unlike many of the Indigenous peoples from Australia and North America, Māori were not legally mandated and/or forcibly removed to live in separate spaces (reservations, reserves, and missions). Yet, just like in other settler-colonial societies, the processes and practices that we now refer to as "planning" in Aotearoa were interwoven with settler-colonial efforts to gain control over territories (Porter, Jackson, and Johnson 2017, 21; Veracini 2010).

The phrases "settle," "settling," "settlement," and "settler" are used deliberately in this chapter, while recognizing that such apolitical and benign-sounding terminology conceals the violence and dispossession of Indigenous people (Porter, Jackson, and Johnson 2017; Veracini 2010). Aotearoa was and still is a settler-colonial society. Settler colonies, unlike extractive colonies designed to exploit natural resources and the colonized labor force, are "predicated on dis-possession and the re-placement of Indigenous people with settlers who do not return to their country of origin" (Porter, Jackson, and Johnson 2017, 3). The settler colonialism of the British Empire, of which Aotearoa was part, involved "large-scale taking and reallocations of land" as well as administrative attention and practices, which continue to constitute the underlying structure and logic of Aotearoa society today. Yet, planning in Aotearoa seldom recognizes the colonial logic that underpins its origins and practices.

This chapter represents just one aspect of Māori planning issues. It brings atten-tion to how a Māori iwi engaged in the emergent planning regime that began to be implemented soon after formal British colonial rule was established in 1840 and became solidified in the first half of the 20th century (Barry and Thompson-Fawcett 2020; Matunga 2013). The author explores the story of one place (Tāmaki Makaurau Auckland, hereafter referred to as Tamaki Makaurau) and one tribal group's (Ngāti Whātau Ōrākei) during the 19th, 20th, and 21st centuries.[1] The example demonstrates how planning was a critical actor and principal structure in the colonization and dispossession of Ngāti Whātau Ōrākei in the city of Tamaki Makaurau. In providing an account of how one Māori iwi engaged with, and was discriminated by, the colonial government and its planning regime, this case also demonstrates the continued presence and ongoing agency of Māori people not only in Tamaki Makaurau's but the entire Aotearoa nation's planning history.

The ongoing intergenerational legacies of Indigenous place-erasure and settler place-making are evident in Aotearoa and in its largest city Tāmaki Makaurau but are being challenged by mana whenua (iwi who hold primary authority over their tribal lands) asserting their rights to, interests in, authority over, and responsibilities toward, their rohe (Thompson-Fawcett, Ruru, and Tipa 2017; Viriaere and Miller 2018). The urban planning canon largely overlooks these Indigenous geographies, which include social and legal actions to maintain Māori land rights, while being complacent about the historical legacies of exclu-sion and subjugation for the most marginalized groups within colonial societies (Barry and Thompson-Fawcett 2020).

1840–1900: a breached treaty between two cultures

Ngāti Whātua is the iwi that held mana (power and sovereignty) over cen-tral Tāmaki Makaurau in 1840 when the Te Tiriti o Waitangi/the Treaty of Waitangi (hereafter the Treaty) was signed.[2] Ngāti Whātua chiefs initially welcomed the arrival of Pākehā (term used by Māori to refer to Europeans) and, following the signing of the Treaty, invited Governor Hobson to create a township on their whenua (AJHR 1879; Orange 2015; Waitangi Tribunal 1991). The iwi made thousands of acres of their lands available for the colonial government (the Crown) to purchase for the new settlement. The costs of estab-lishing the city of Tāmaki Makaurau – which was later to become Auckland,

Aotearoa's largest city – were funded from the vast profits the Crown generated by purchasing land off Ngāti Whātua Māori at a low price and then on-selling the land at a higher price to settlers. The iwi and the Crown agreed to the iwi being given a 10% resale profit after the first lot of land sales. However, historians concur, the majority of that money was spent by colonial authorities on schools, hospitals, and roads for the entire city population rather than on works that directly benefited Ngāti Whātua (Anderson, Binney, and Harris 2015; Waitangi Tribunal 1991).

Ngāti Whātau wanted to maintain their rangatiratanga, mana, and tikanga, but also wanted to ensure that they created harmonious relations with Pākehā. As part of this relationship-building, which rested at the heart of the Treaty, the iwi gifted some of its lands at Ōrākei (located beside the harbor near the center of the city) to the Anglican Church for the purpose of establishing a school and a chapel. Iwi elders wanted their whānau (family) educated in both Māori and Pākehā ways. Likewise, in 1859, when the British Colonial Office and its government representatives became convinced that the Russians were looking to invade Aotearoa, Ngāti Whātau gave a section of Ōrākei headland for defense purposes (Waitangi Tribunal 1991). In line with tikanga (Māori laws), Ngāti Whātau gave both parcels of land to the Anglican Church and the Crown as gifts that were meant to be returned to the iwi when both groups no longer required the whenua for education and defense purposes; this was not done, which breached the tikanga of the iwi (Waitangi Tribunal 1991).

Ngāti Whātua continued to support the Crown and the embryonic city in the 1860s and 1870s when warfare between various iwi and the Crown erupted throughout the North Island (known as the New Zealand Wars). This stood in marked contrast to many other iwi, including their southern neighbors of Waikato-Tainui, Ngāti Maniapoto, Raukawa, Te Arawa, and Ngāti Tuawharetoa. Later, Ngāti Whātau established a Māori Parliament at Ōrākei and emphasized their role as leaders within the Māori world, as well as the business, civil and political affairs of the settler-colonial colony. Ngāti Whātau leadership also created iwi policies and actively supported Pākehā within Tāmaki Makaurau (Anderson, Binney, and Harris 2015). Yet, they always insisted on their and other tribes' rights to self-determinate their own affairs (rangatiratanga) (AJHR 1879).

By the end of the 19th century, Ngāti Whātua had sold the vast majority of their land to the Crown – except the tribe's main base at Ōrākei, which was only 700 acres in size (Waitangi Tribunal 1991). Ngāti Whātua made it clear that the Ōrākei land block was special to them for socioeconomic, cultural and spiritual reasons, and they wanted to ensure it remained in their possession permanently (which was something guaranteed to them under the Treaty). Despite Aotearoa Governor Grey's assurances to Te Kawau (iwi's pre-eminent rangatira), the Crown already began to pass legislation and establish colonial institutions designed to dispossess iwi, including Ngāti Whātau, of their remaining lands.

Historian Binney (2001) likens the Native Lands Acts of 1862 and 1865 (New Zealand Parliament 1862, 1865), which created and regulated the

Native Land Court (NLC), to an act of war. Their operations were of calamitous effect on Ngāti Whātau. The Court required that all land held by Māori be converted to Western-style titles. Each title to a land block (irrespective of its size) was allowed to have no more than ten owners; this was later changed in 1873, so there was no restriction on the number of owners. Owners could use, lease, and sell their lands without the consent of the wider iwi. All the other hapū and iwi members were legally dispossessed of their rights. The expenses from gaining land title resulted in many Māori selling their lands to pay the costs incurred (survey costs often were 20%–25% of the total value of a block).

At Ōrākei, the impacts of the NLC were particularly disastrous, with Māori dispossession a consequence of the imposition of individual land titles along-side capitalism and settler-colonial sociopolitical structures. In 1869, the Court awarded the entire block of land (700 acres) to thirteen individual iwi members (who were either chiefs themselves or men of rank who were related to one of three Ngāti Whātau chiefs). The more than 300 Ngati Whātau people who lived at Ōrākei as well as others living elsewhere within the region were legally disin-herited. The decision of who was to be named on the title was made at a meeting held in early 1869, which was attended by representatives from three prominent chiefs (Te Kawau, Wiremu Watene, and Arama Te Karaka) and their Pākehā lawyers. No historical records survive from that meeting and it remains unclear if any wahine (women) attended the meeting.

At the time of colonization Māori wahine throughout the country enjoyed far greater political and economic freedoms than their counterparts in Victorian Britain. For instance, kahurangi (chieftainess) and wahine rangatira (women of rank) were able to inherit exclusive land rights from their mothers and occupy positions of polit-ical, social and economic influence within their hāpu and iwi (Parsons, Fisher, and Crease 2021). However, colonial officials sought to impose Victorian British gender norms on Aotearoa, which included actions to restrict the rights of Māori wahine; this included efforts to prevent women from signing the Treaty in 1840, as well as being named on land titles in the 1860s–1890s (Brookes 2016).

In 1869, most iwi members were not concerned with land titles as they believed that the owners were similar to trustees (Waitangi Tribunal 2001); their view was incorrect according to settler-colonial law but correct according to their tikanga. Individual people being named as owners of particular parcels of land was diametrically opposed to Te Ao Māori (the Māori world or Māori worldviews), wherein land could not be owned (as it was both an ancestor and a member of one's kin group). Māori land tenure arrangements, in contrast to those of Britain and other Western cultures, centered on rights to use land, water, and other resources, as well as on duties of care (kaitiakitanga/environ-mental guardianship). Iwi and hapū shared overlapping rights and responsibilities to particular lands and waters, but some held more authority than others (known as mana whenua) (Parsons, Fisher, and Crease 2021).

Since no individual or group could own land, the concepts of ownership, land titles, and land sales were incomprehensible to most Māori throughout the

19th century. Accordingly, the creation of the NLC (alongside ongoing land sales) represented a radical change for Māori, who were forced to grapple with the ontological and epistemological differences embedded in the emergent settler-colonial society and its legal and planning regimes. The NLC operations deliberately undermined the customary governance arrangements and planning practices of Ngāti Whātau (and another iwi) (Binney 2001).

The leadership of Ngāti Whātau, led by chief Tuhaere, continued to engage with government officials and colonial institutions in the hope of ensuring that their kin was able to adapt and live well in the rapidly growing and changing city of Tāmaki Makaurau. Tuhaere, for instance, arranged for a special Act of Parliament to allow him to undertake a subdivision at Bastion Point (opposite the land his iwi gifted to the Crown for defense purposes) (Clark 2000). Under his leadership, the rental payments derived from the lands the iwi leased to Pākehā were funneled into schemes to maintain the homes and livelihoods of Ngāti Whātua. This is an example of how chiefs sought to maintain their rangatiratanga (chiefly authority) over their whenua. However, the efforts of Ngāti Whātua were undermined by the ongoing activities of the NLC.

In 1898, the NLC divided the whole of the Ōrākei land block into even smaller land parcels (to individual titleholders) which effectively ended any notion of iwi control (Waitangi Tribunal 1991). For Ngāti Whātua, it was made clear by this act that the thirteen people (their kin) who were named on the original title to Ōrākei block were not trustees (in line with their tikanga) but were in fact owners (as per the British legal tradition). Through this action, the Court proclaimed that the settler-colonial legal system took precedent over tikanga Māori.

Unsurprisingly, Ngāti Whātua protested about the Court's decision (AJHR 1871, 1879). They mounted legal actions and took a petition to the colonial parliament, which aimed to get their own land tenure arrangements restored or, if this was unachievable, for everyone in the iwi to be included as owners on the land parcels. All the protests were unsuccessful. Many individual owners, outside the customary governance structures and management approaches of the iwi, decided to lease out their land to Pākehā.

1900–1945: planning Auckland suburbs while further dispossessing Māori

By the start of the 20th century, Ngāti Whātua were in financially difficult situations without enough land to grow cultivations or farm livestock, lacking access to finances, and limited job opportunities. In short, they were socioeconomically deprived while all around them, the suburbs of Ōrākei, Remuera, and Parnell were becoming filled with wealthy Pākehā residents. In 1910, Pākehā who leased land (from individual owners) at Ōrākei petitioned parliament and demanded that they be given the right to buy their properties as freehold titles (Waitangi Tribunal 1991). The leaseholders argued that they had made significant improvements to the land (building houses and creating attractive gardens) and they should be the ones who received the financial benefits from their hard work (not the Māori landowners) (Waitangi Tribunal 1991). Their argument tapped into a longstanding settler-colonial narrative which held that Indigenous people did not "use" their lands (left them as "wastelands" or "wilderness"); a narrative that also paralleled Weber's Protestant work ethic that declared that only those who worked the land deserved to own it. In other words, Indigenous dispossession was justified on the basis that settlers did use the lands in *profitable* manner (clearing, developing, farming, etc.) (Veracini 2010).

Ultimately, the local government decided on another tactic: it convinced the city's Pākehā parliamentarians to introduce a Bill into Parliament that would see the Crown compulsory acquire all the Māori land at Ōrākei and develop it for residential and recreational purposes (except for the Ōkaku Bay kainga/village) (Unknown Author 1909b, 1912, 1929). Once again, Ngāti Whātua campaigned against the proposed Bill (Unknown Author 1939; Waitangi Tribunal 1991) but their protests ultimately failed with both the courts and parliament largely ignoring them. Although they received support from the Native Affairs Committee of the House of Representatives, this support did not translate into on-the-ground actions.

At the same time, the government introduced a special act of parliament to compulsory acquire a small parcel of Ngāti Whātua land to construct a sewerage line (Waitangi Tribunal 1991). It was the iwi's first experience but not their last of compulsory acquisition (also known as resumption, expropriation, or eminent domain). The pipeline diminished the accessibility of the beach and harbor for Ngāti Whātua iwi members who lived in the kainga. The construction of the wider sewage system at Ōrākei, located inland toward the hill, also involved the removal of vegetation, drainage of wetlands, and realignment of streams, all of which resulted in the disruption of the natural flow of water. Subsequently, Tāmaki Makaurau residents reported that the land at Ōkaku Bay was increasingly swampy and prone to flooding as a consequence of the sewage works (Unknown Author 1931, 1937b, 1940). The sewage polluted the mahangi kai (food gathering sites) of the iwi. This caused a serious health threat for Māori more than others as they relied on food harvested from the bay and also swam in its waters. There were reports of Māori becoming seriously ill and some even dying from typhoid as a consequence of eating shellfish contaminated by untreated sewage. Rather than inspiring sympathy, this was used as evidence by government officials as well as Pākehā residents that Māori modes of living were unhealthy, and they should be removed from the area to ensure the safety of the wider public (specifically Pākehā) (Unknown Author 1928, 1932, 1940). As a consequence of this act, as well as other pressures, many Māori moved away from their village. This experience paralleled the experiences of Māori elsewhere in the country: they were exposed more frequently to environmental risks than Pākehā due to environmental racism.

After some of the individual Māori landowners sold their titles to leaseholders in 1912, the Crown announced its intention to buy the whole of Ōrākei. Central and local government officials alongside Pākehā residents within the area deemed Māori houses at Ōrākei to be "old shacks" and a "slum"

(Unknown Author 1937a, 1943, 1950b). Ngāti Whātua wanted to retain the land and resisted compulsory acquisition in various ways; those without any land titles moved to church land (until that too was acquired by the Crown in 1926). The government's efforts to acquire all interests in the Ōrākei took more than thirty years (1913–1950). Throughout this period, the local government asserted ongoing pressure on the British Crown to acquire (i.e., dispossess) all the remaining lands of Ngāti Whātua and forcibly remove those iwi members who wanted to stay living within their kainga.

By the 1930s, the majority of the land had been purchased, and the government set out its plan to develop a model garden suburb. The suburb of Ōrākei was the centerpiece of the first Labour Government's state housing plan for the city of Tāmaki Makaurau (Miller 2004; Unknown Author 1924). It was designed with individual houses on good-sized sections, large public green spaces, scenic views, and curving streets, which were all exceptions to how suburbs were designed in Aotearoa, where the focus was on practical, economical, and fast suburban growth (Miller 2004; Unknown Author 1924). More land was taken by the Crown for defense purposes during WWII, with the whānau and hapū of Ngāti Whātua who remained living in Ōrākei forced to live in an ever-decreasing area (Waitangi Tribunal 1991). However, even this small landmass was highly prized by the local government (Auckland City Council), the Crown, and property developers due to its sea views and close proximity to the city's Central Business District (Clark 2000).

The government's efforts to purchase land in Ōrākei were regularly challenged and resisted by Ngāti Whātua, who consistently spoke out in opposition to the government's purchasing and development plans for their land. Numerous legal cases were filed by representatives of Ngāti Whātua in the Native Land Court, Supreme Court, Court of Appeal, and Compensation Court, as well as appearances before the Commission of Inquiry and Parliamentary Petitions (Unknown Author 1939). All these efforts were from an iwi who were labeled by the Crown as "willing sellers." In reality, a wealth of historical evidence demonstrates that many Ngāti Whātau opposed land sales. Iwi members who retained land and sold it to the Crown often did so under duress or in the hope that if they sold their plot of land, then other parts of the village would be protected (Unknown Author 1935, 1939, 1944a).

1945–1965: growth of urban Māori and racial discrimination in social housing

In 1950, the Crown once again used their powers of compulsory acquisition to take more Māori land in Ōrākei, which included the village and marae (a fenced-in complex of carved buildings and grounds of a particular hapū/subtribe) at Ōkaku Bay. The entirety of the village, where many Ngāti Whātau whānau continued to live, was acquired under the Public Works Act for the purpose of demolishing the buildings and turning them into a city park. The newspaper reported the Crown's difficulties in removing Māori living at Ōrākei in overcrowded conditions in "shacks, tents, and an old house," and how many refused to leave (Unknown Author 1950b) even once the Crown took legal possession of the area (approximately two and a half acres). However, in 1951–1952 the government forcibly evicted the remaining 35 Ngāti Whātau households. Iwi members were forced to watch as their homes and marae were dismantled and burnt to the ground by the Crown officials, and the area hastily converted into a park. The households were relocated (removed) to elsewhere to live as tenants of government housing within Ōrākei as well as elsewhere in the city and beyond its boundaries.

In other parts of Aotearoa, Māori were likewise experiencing the negative impacts of Crown policies and practices concerning how the land was planned, allocated, and used. In particular, the operations of the NLC resulted in many Māori whānau, hapū and iwi being left with little or no land to cultivate, farm, or log (activities that had been the mainstays of the Māori economy). Land dispossession, lack of job opportunities in rural areas, and wider sociopolitical discrimination all contributed to Māori migration to the cities. Roughly 30% of the Māori population lived in Aotearoa's urban areas in the mid-20th century. By 1966, this number had more than doubled (66%), and by the end of the 20th century, nearly 85% of Māori were urban dwellers (Hill 2012). The majority of Māori migrated to Tāmaki Makaurau, and they became mātāwaka – i.e., Māori who did not hold mana (power) or whakapapa (genealogical) links to the land on which they lived. Instead, mana whenua status remained within the various iwi (including Ngāti Whātua, Waikato-Tainui, Te Kawerau ā Maki).

Māori who moved to Tāmaki Makaurau for education or employment opportunities frequently encountered severe levels of poverty and discrimination, which paralleled the experiences of Ngāti Whātua and other mana whenua of the city. Māori were discriminated against in numerous ways by residents, businesses, and government agencies in a city where Pākehā made up the largest ethnic group (Hill 2012). Māori recall their experiences of hotels refusing to allow them to stay on their premises, shopkeepers denying them service, and banks refusing their application for loans, all on the basis of their ethnicity (Bartholomew 2020, 98). This was never an official policy (unlike in South Africa or the southern states of the United States), but rather an unwritten rule which highlights how institutional racism was embedded throughout Aotearoa settler-colonial society.

Māori also encountered racial discrimination when trying to find houses or rooms to rent in Tāmaki Makaurau. Although government officials rejected Māori complaints about discrimination, a journalist published a series of articles in Auckland newspapers in 1953 which exposed the discrimination (Bartholomew 2020; Taylor 1953). One Māori woman whom Taylor interviewed described the difficulties she faced finding accommodation in the 1950s, with accommodation agencies telling her: "No, it's no use you … coming down [to visit our office]. We can't place Maoris [sic] … References won't help you; it's the colour … Not many of our clients take Maoris [sic]" (Taylor 1953, 4). In a wider context in which racism was socially and politically normalized, planning and planners' exclusion of Māori perspectives was in many ways unsurprising.

In the mid-20th century, Aotearoa government policies were implicitly premised on the idea of cultural assimilation, which paralleled those promoted in some settler colonial states/provinces in Australia, United States, and Canada (Ellinghaus 2006). Cultural assimilationist ideas (Ellinghaus 2006) held that Indigenous people must adopt white (Pākehā) ways of dressing, speaking, eating and living in order to fit into the new settler-colonial order of things and become "civilized subjects."

The Māori Social and Economic Advancement Act of 1945, for instance, was "designed to facilitate the full integration of the Māori race into the social and economic structure of the country" (New Zealand Parliament 1945). Social welfare, including access to government housing, was linked to the capacities of Māori to become successful inhabitants of Pākehā society through integration and cultural assimilation. According to the New Zealand Parliament, "an important feature of the [1945] Act is that it does not seek to impose standards from without; rather it calls upon the Maori people to exercise control and direction of their own communities in the essentials of good citizenship and responsibility" (New Zealand Parliament 1945). The vagueness of the definitions of "standards," "good citizenship," and "responsibility" meant that much was left up to government officials (most of whom were Pākehā) to interpret whether and how Māori were demonstrating suitable evidence of social and economic advancement.

Initially, the state housing scheme, which began in the early 1930s under the first Labour Government, only permitted Pākehā families to be tenants of state houses; this meant that no Māori were eligible to live in the Ōrākei garden suburb. It was not until 1948 that low-income Māori nuclear families were finally eligible for state houses. However, statistics released in 1949 reveal that of the 30,000 state houses allocated to families throughout Aotearoa, only 100 were Māori (Carlyon 2013). This ratio of one Māori family receiving a state house to 300 Pākehā/European families was significantly lower than the demographic composition of Aotearoa at the time; by 1940 the ratio Māori to Pākehā was roughly 5.5 to 100.

In addition, the government also instituted a policy of "pepper-potting" Māori families among Pākehā neighbors. What this meant in practice was that a single Māori family would be given a state house (typically a three-bedroom home) on the street surrounded by Pākehā. The domestic metaphor of pepper is striking and suggests a process whereby "black" pepper is diluted in "white" flour to ensure it is not too visible or overwhelming (Labrum 2013). In reality, however, Māori (the "pepper") was noticeable and Pākehā residents frequently wrote to government officials and local newspapers to complain about the "Māori problem" in the city (Unknown Author 1909a, 1937a, 1943, 1950a). Complaints by members of the public (paralleling those of government officials) included that Māori houses were overcrowded, unhygienic, and poorly maintained, which diminished the quality of the entire neighborhood (Labrum 2013). Pākehā government officials ascribed Māori people's "appalling" living conditions (in dirt-floored "slums") to their passivity and laziness rather than acknowledging the institutional racism they faced (Unknown Author 1944b, 7).

The government's planned dispersal of the Māori population (across different streets, suburbs and cities) was designed to ensure that Māori integrated (assimilated) into Pākehā culture. Just as Māori children were prohibited from speaking the Māori language in schools, the government sought to "encourage" (or force) Māori to shift away from their communal living arrangements (revolving around kinship ties) to those that mimicked Pākehā. When Māori received a state house, they were expected to live as a single-family unit (husband and wife plus their children). Government officials came to evaluate the behavior of Māori households (under the guise of health and hygiene inspections). Those who allowed their Māori kin to stay with them were deemed troublesome tenants who failed to conform to expected standards of hygiene and (Pākehā) social decorum (Labrum 2013).

Yet, the nuclear family was a concept foreign to Māori, as whānau (family) was understood to include wider kinship groupings including grandparents, aunties, uncles, and cousins. Rather than assimilating into Pākehā culture, the oral histories of Māori who moved to the city reveal that Māori adapted and preserved many aspects of their Māori selves and Māori ways-of-being including notions of home in urban areas. The marae was (and still is) the traditional spatial formation for everyday communal living for Māori, based on kinship systems of whānau, hapū, and iwi affiliations, and consisted of physical structures for sleeping, meeting, food, and ablution. Since Māori who migrated to urban areas lacked a marae, Māori recount in their oral histories how their house (or their auntie's, grandmother's, etc.) became the key meeting place for everyone in their whānau. As King et al. (2018, p. 1198) write: "The significant feature of such spaces [was] that they embrac[ed] Māori values of manakitanga (care and hosting others) and whanaungatanga (kinship, sense of familial connections), and offer[ed] locales that support[ed] Māori ways-of-being" within Pākehā cityscapes. Māori who lived in the city sought to maintain their cultural identities through simple acts like searching for and locating, then cooking and sharing culturally significant foodstuffs with their whānau and friends. Rather than being isolated from Māori culture and being forced to assimilate into Pākehā culture, Māori historians highlight how urban Māori retained strong ties to their iwi (to those back in their home communities as well as those living in urban areas) (King et al. 2018; Matutina Williams 2015).

1965–1980: Ngāti Whātau efforts to protect their lands and resist unjust planning decisions

This was a time of crisis in planning and race relations in Aotearoa (Harris 2004). In 1976, the Crown proposed to sell the public reserve at Bastion Point, which located overlooked Ōkaku Bay, for a luxury housing development (Harris 2004;

Hayden 2018). Ngāti Whātau had given the land (known as Takaparawhā) to the Crown in 1885 for defense purposes (against the supposed threat of Russian invasion). The Crown had converted Bastion Point to a reserve (Savage Memorial Park) that commemorated the burial place of the first Labour Prime Minister (Michael Joseph Savage), who died in 1940.

However, Ngāti Whātau had long desired and campaigned for the land at Bastion Point to be returned to them. They argued that, as per tikanga Māori, their ancestor (chief Tuhaere) gave the land as a gift to the Crown to be used for national defense and, in line with tikanga, the gift was not a permanent transfer of ownership but rather a temporary usage right. Yet, the Crown had failed to return it (Clark 2000; Waitangi Tribunal 1991).

On 5 January 1977, only two days before the official building work was meant to start, a group of Ngāti Whātua (under the name of Ōrākei Māori Action Committee) began their occupation of Bastion Point (Harris 2004). The protesters, who were supported by hundreds of other Māori from different iwi, occupied the site for 506 days. The protest ended when the government sent in the police and army to forcibly remove 222 Māori people. Their temporary marae, building, and gardens were demolished, and basic amenities were withheld from the site to ensure that no one returned. The new housing development at Bastion Point, however, was never constructed.

1980s–2000s: recognition of Māori rights in a neoliberal planning milieu

In the mid-1980s, the Waitangi Tribunal began to investigate Ngāti Whātau's claims that the Crown failed to honor the articles of the Treaty with regard to the Ōrākei land block. The Tribunal had been established in 1975 as a permanent commission of inquiry to investigate any Māori claims of the Crown's Treaty breaches. It released its findings in 1991, which upheld Ngāti Whātau's argument, and recommended that the land be returned to Ngāti Whātau. The iwi and the Crown eventually agreed to a Treaty Settlement; this included a formal apology from the Crown for past injustices against the iwi as well as financial compensation and the return of public land at Bastion Point (used as parks and state housing) (Waitangi Tribunal 1991).[3] A later Treaty Settlement between Ngāti Whātua and the Crown saw the return of land elsewhere in the city and provided a larger sum of money to compensate iwi for their experiences of dispossession, discrimination, and deprivation over one hundred plus years. The Treaty settlements allowed Ngāti Whātau to begin to directly take action to address the litany of social and environmental injustices they experienced, which was all the more critical since the advent of neoliberalism in Aotearoa.

From the mid-1980s onwards successive national governments introduced neoliberal policies (promoting economic growth and reducing government expenditure and controls). State houses were sold off, the building and construction sectors were deregulated, and government spending on infrastructure and social welfare reduced (Brookes 2016). Neoliberal reforms resulted in heightened socioeconomic deprivation, particularly among Māori, and loss of secure housing as the housing market became privatized and increasingly unaffordable. Planning became a tool to sustain capital accumulation in neoliberal Aotearoa with the planning interests of higher income groups well catered for but those of middle- and lower-income groups (including many Māori) largely ignored. It is within this setting that Ngāti Whātau sought to plan for and enact strategies to address the social, economic, and ecological needs of its iwi members.

2010s–present: decolonizing planning in the post-Treaty Settlement era

The Treaty Settlements allowed Ngāti Whātua greater capacities to express and exercise their decision-making rights as mana whenua by providing them with legal recognition of their status and financial resources. These resources started being drawn on to fund Indigenous-led development and ecological restoration projects (Walker et al. 2019), as well as mounting legal challenges against government- or private developer-led plans. The commercial arm of Ngāti Whātau started work on mixed residential and commercial developments in the North Shore (across the harbor from Ōrākei). The iwi was able to sell houses at a profit and use the money to fund the building of low-rent housing for its members (specifically the elderly).

Initially, the iwi's development project was opposed by some residents (comprised chiefly of wealthy Pākehā) who felt the iwi-led development would decrease the value of their properties and contribute to anti-social behavior (Unknown Author 2017). Such NIMBY perspectives rarely use highly racialized words explicitly. Yet, they continue to be coded in the language of settler-colonial privilege which holds that Māori (and other non-whites) are a disruptive and potentially dangerous threat to the Pākehā ways-of-being.

Iwi were forced to fight through planning processes and, in some instances, mounting legal cases to allow them to develop their lands (which they fought so hard to reclaim). Ultimately, they were able to implement the development projects, which were declared a wholesale success: they generated high profits for the iwi corporation and allowed it the construct good-quality affordable housing for iwi members.

The Ngāti Whātua iwi continues to exercise their rangatiratanga by engaging with and sometimes seeking to challenge the Crown, the local government (Auckland Council), and private companies through settler-colonial constructed planning and juridical processes. Some of these involve productive alliances between Māori and non-Māori actors (possessing different types of knowledge) united in a common goal. For example, planners and scientists have collaborated with iwi to restore riparian vegetation and wetlands within the city (Walker et al. 2019). Other plans and strategies are wholly iwi-driven and controlled. These include efforts to create affordable and good-quality housing for elderly iwi members, new economic development opportunities, and healthy landscapes at Ōrākei. Iwi members, for instance, are creating communal food gardens to improve people's access to healthy fruit and vegetables (Viriaere and Miller 2018).

These examples notwithstanding, planner Livesey (2019) concludes that planning in Aotearoa remains a sociopolitical and cultural practice still thoroughly situated within settler-colonial ways of thinking, doing, and organizing spaces. Even when some of their whenua is returned to them, the legal cases indicate that the iwi still face the insidious and persistent processes of settler-colonialism, which continue to function in a way that upholds the planning regulations and socioeconomic structures designed by and for settler-colonial society. Under these circumstances, Māori interests, knowledge, values and the Treaty are delegitimatized or largely irrelevant to the decisions about the planning and management of spaces. Yet, Māori arguably possess greater opportunities to participate in and shape planning processes as a consequence of the Treaty Settlements, a national parliament than contains growing numbers of Māori members of parliament, as well as the fact Māori are a sizeable minority within Aotearoa (making up 15% of the total population).[4]

Conclusion

This chapter demonstrates how settler-colonial planning regimes (both in the past and the present) are not neutral institutions or mechanisms but instead shaped by the sociopolitical contexts in which they operate. As Matunga (2013, 4) observes "planning is not just a word. It is also an imperial scholarly discipline." Recent efforts of governments to acknowledge past wrongs and centuries of social and environmental injustices experienced by Indigenous peoples are insufficient when directly solely at the payment of money and the return of resources (Livesey 2019). Just and equitable planning processes within settler-colonial societies, as Matunga (2000) suggests, requires the (re)establishment of Māori authority within planning processes and Māori-led developments. Some go so far as to suggest that decolonizing planning necessitates Indigenous-led planning processes (or complete Indigenous autonomy). The historical case of Ngāti Whātau Ōrākei, presented in this chapter, highlights the critical need for scholars and practitioners to not only recognize Māori cultural identities and relationships with their whenua, but also to critically evaluate the ways in which planning structures and mechanisms operate. Indeed, there is a need to consider if and how Te Ao Māori is being incorporated and empowered within legislation, regulations, and procedures governing planning practices. How can a decolonizing praxis be adopted to ensure that Māori planning interests, needs, and priorities can be addressed in an equitable, effective, and tikanga-appropriate manner?

Alternative planning history and theory timeline

Period	Planning history and theory canon	Alternative timeline: Indigenous people
Late 1800s–1900	Birth of planning	A breached Treaty between two cultures Signing of Treaty of Waitangi, establishing a city (Tāmaki Makaurau Auckland), legally dispossessing Māori
1900–1945	Formalization of planning	Planning Auckland suburbs while further dispossessing Māori Compulsory land acquisitions by the Crown, first instances of environmental racism, first garden suburb in Ōrākei
1945–1965	Growth of planning	Growth of urban Māori and racial discrimination in social housing Māori mass migration to cities, poverty and discrimination in housing and public space, "pepper-potting" Māori families among Pākehā neighbors, denial of Māori lifestyle traditions
1965–1980	Midlife crisis of planning	Ngāti Whātau efforts to protect their lands and resist unjust planning decisions Māori occupation of Bastion Point for 506 days, Māori succeed in blocking luxury housing development at Bastion Point
1980–2000	Maturation of planning	Recognition of Māori rights in a neoliberal planning milieu Waitangi Tribunal finds in favor of Māori, Crown agrees to Treaty Settlement, land returned to iwi and formal apology issued, neoliberal planning undermines Māori efforts
2010–present	New planning crisis	Decolonizing planning in the post-Treaty Settlement era Indigenous-led development and ecological restoration projects, legal challenges, fighting Pākehā NIMBYism, planning remains a settler-colonial practice

Case study: Landmark court decisions in favor of Ngāti Whātau

Ngāti Whātau continue to both negotiate with, and challenge, settler-colonial legal and planning systems in instances where their tikanga, values, and status as mana whenua are ignored by governments, planners and/or developers.

Ngāti Whātua Ōrākei, for instance, filed a case with the Environment Court in the late 2010s to oppose a marina development project that was approved by the local government. This was funded by another iwi – Ngāti Maru – whose lands are located south in the Taranaki region. Ngāti Whātua lodged the legal case on the basis that, while all iwi within the Tāmaki Makaurau region were consulted, their status as mana whenua (ultimate holders of tribal authority) was not given priority, and their history, knowledge and tikanga were disregarded. This breached the terms of the 1991 Resource Management Act as well as the Council's own mana whenua engagement policy (New Zealand Parliament 1991). In 2020, the Environment Court and later the High Court found in favor of Ngāti Whātua's claims (Whata 2020). Legal scholars declare that the Environment Court's and High Court's decisions are unprecedented in terms of planning case law in Aotearoa. For the first time, the courts found that certain iwi possess greater authority (mana) than others and that both governments and developers are required to identify who the mana whenua group is and devote their time and resources toward meaningfully consulting with, and gaining the support of, mana whenua for new plans and projects. It means that governments' and developers' efforts to simply engage with all iwi groups in a location are insufficient to ensure that mātauaranga Māori, tikanga Māori, and rangatira-tanga are recognized and taken into account in planning decisions.

Notes

1 It is important to acknowledge that while the author is a Māori researcher, their iwi is Ngāipuhi (whose rohe is located in the upper northland peninsula of Aotearoa's North Island). Therefore the author is writing about the history of another iwi. The author's whānau (family) did, however, migrate to Tamaki Makaurau more than a century ago, after being dispossessed of their lands, and lived within the rohe of Ngāti Whātau. Many were first-hand witnesses to many of the events described in this chapter. Thus, while this is not the author's history, there are personal connections to the story of how Ngāti Whātau sought to retain their lands in the face of ever more intrusive and discriminatory government laws and planning practices. Thus, the author would like to acknowledge that they and their whānau are mātāwaka (Māori without genealogical connections) of the city of Tāmaki Makaurau and always need to acknowledge the rangatiratanga (chiefly authority) of the mana whenua groups (tribal groups with supreme authority over their rohe).

2 The formal British colonial rule of Aotearoa began after the signing of Te Tiriti o Waitangi (also known as the Treaty of Waitangi or the Treaty); in 1840 various copies of the Treaty (written in English or Māori) were signed by representatives of the British government and 500 Māori rangatira/chiefs. Rangatira who signed the Treaty signed the Māori version, whereas British signed the English version. There were substantive differences in the wording (and meaning) of the English and Māori versions. The Māori version emphasized the Māori would retain their sovereignty and decision-making authority their lands/waters/resources, while agreeing to Britain becoming partners with Māori and be involved in governing non-Māori residents in Aotearoa and international relationships (the Māori word governorship was used to describe this relationship). Meanwhile, the English version outlined that Māori ceded their sovereignty to Britain but would retain their ownership of land and other resources in return for gaining the rights of British subjects. Not all Māori signed the

Treaty, and even those who did assumed it meant they would retain their sovereignty and the document involved the development of a partnership with Britain (nation-to-nation of equal partners working together). The British Government disagreed on the principles of the Treaty (both English and Māori versions) and colonial violence and dispossession of Māori began to occur soon after its signing (Orange 2015).

3 A Treaty Settlement is a legal agreement, a practice that commenced in the mid-1990s, by which the New Zealand Crown (central Government) reaches an agreement with an individual iwi regarding how to compensate iwi for their historical and contemporary experiences of dispossession, violence, discrimination and injustice (as it related to specific breaches of the articles of the Treaty). Treaty Settlements typically include the Crown issuing a formal apology to the iwi for historical and contemporary injustices, providing a financial compensation package (money, return of land, forests), right of first purchase on Crown-owned assets, introduction of new legislation, and more recently (since the mid-2000s) the establishment of co-governance and co-management arrangements over rivers, lakes or forests (Mutu 2019; Wheen and Hayward 2012). Ngāti Whātau's Treaty settlement did not include the Savage Memorial Reserve at Ōrākei (since it was the burial site/urupā of former Prime Minister Michael Joseph Savage and therefore considered tapu/sacred).

4 Although the Māori population demographically declined in the 19th century as a consequence of colonial activities (warfare, land takings, introduction of infectious diseases), the population decline was less extreme than elsewhere (largely due to the absence of smallpox, which discriminated Australian and American Indigenous populations). Although settlers had declared Māori a "dying race" in the 19th century, Māori population began to recover in the 20th century as a consequence of the end of warfare, improved public health and community infrastructure, as well as a high birth rate. By the start of the 21st century Māori comprised 15% of Aotearoa's population. In contrast, in other settler-colonial societies Indigenous peoples' make up less than 5% of the total population (3.3% in Australia, 2% in the United States, and 4.9% in Canada).

References

AJHR. 1871. "Petition of Paora Tuhaere." *Appendix to the Journal of the House of Representatives*. Wellington: Government Printer.

———. 1879. "Paora Tuhaere's Parliament at Orakei." *Appendix to the Journal of the House of Representatives*. Wellington: Government Printer.

Anderson, Atholl, Judith Binney, and Aroha Harris. 2015. *Tangata Whenua: A History*. Wellington: Bridget Williams Books.

Barry, Janice, and Michelle Thompson-Fawcett. 2020. "Decolonizing the Boundaries between the 'Planner' and the 'Planned': Implications of Indigenous Property Development." *Planning Theory & Practice* 21 (3): 410–25.

Bartholomew, Robert E. 2020. *No Māori Allowed: New Zealand's Forgotten History of Racial Segregation*. Auckland: BookPrint.

Binney, Judith. 2001. "The Native Land Court and the Maori Communities." In *The Shaping of History: Essays from the New Zealand Journal of History*, edited by Judith Binney, Judith Bassett, and Eric Olssen, 143. Wellington: Bridget Williams Books.

Brookes, Barbara. 2016. *A History of New Zealand Women*. Wellington: Bridget Williams Books.

Byrnes, Giselle. 2001. *Boundary Markers: Land Surveying and the Colonisation of New Zealand*. Wellington: Bridget Williams Books.

Carlyon, Jenny. 2013. *Changing Times: New Zealand since 1945*. Auckland: Auckland University Press.

Clark, Precious. 2000. "Te Mana Whenua O Ngati Whatua O Orakei." *Auckland University Law Review* 9: 562.

Ellinghaus, Katherine. 2006. "Indigenous Assimilation and Absorption in the United States and Australia." *Pacific Historical Review* 75 (4): 563–85.

Harris, Aroha. 2004. *Hīkoi: Forty Years of Māori Protest.* Wellington: Huia Publishers.

Hayden, Leonie. 2018. "The Occupation of Takaparawhā Bastion Point, 40 Years on." *The Spinoff.* May 26, 2018. https://thespinoff.co.nz/atea/26-05-2018/40-years-on-from-the-occupation-of-takaparawha-bastion-point/.

Hill, Richard S. 2012. "Maori Urban Migration and the Assertion of Indigeneity in Aotearoa/New Zealand, 1945–1975." *Interventions* 14 (2): 256–78.

King, Pita, Darrin Hodgetts, Mohi Rua, and Mandy Morgan. 2018. "When the Marae Moves into the City: Being Māori in Urban Palmerston North." *City & Community* 17 (4): 1189–208.

Labrum, Bronwyn. 2013. "Not on Our Street: New Urban Spaces of Interracial Intimacy in 1950s and 1960s New Zealand." *Journal of New Zealand Studies* (14): 67–86.

Livesey, Brigid Te Ao McCallum. 2019. "'Returning Resources Alone Is Not Enough': Imagining Urban Planning after Treaty Settlements in Aotearoa New Zealand." *Settler Colonial Studies* 9 (2): 266–83.

Matunga, H. (2000). Decolonising planning: The Treaty of Waitangi, the environment and a dual planning tradition. Environmental Planning and Management in New Zealand, 36–47.

Matunga, Hirini. 2013. "Theorizing Indigenous Planning." In *Reclaiming Indigenous Planning*, edited by David C. Natcher, Ryan Christopher Walker, and Theodore S. Jojola, 1–32. Montreal & Kingston: McGill-Queen's University Press.

Matutina Williams, Melissa. 2015. *Panguru and the City: Kāinga Tahi, Kāinga Rua, An Urban Migration History.* 1st ed. Wellington: Bridget Williams Books.

Miller, Caroline. 2004. "Theory Poorly Practised: The Garden Suburb in New Zealand." *Planning Perspectives* 19 (1): 37–55.

Mutu, M. (2019). The Treaty Claims Settlement Process in New Zealand and Its Impact on Māori. Land, 8(10), 152. https://doi.org/10.3390/land8100152

New Zealand Parliament. 1862. *Native Lands Act.*

———. 1865. *Native Lands Act.*

———. 1945. *Māori Social and Economic Advancement Act.*

———. 1991. *Resource Management Act.*

Orange, Claudia. 2015. *The Treaty of Waitangi.* Wellington: Bridget Williams Books.

Parsons, Meg, Karen Fisher, and Roa Petra Crease. 2021. *Decolonising Blue Spaces in the Anthropocene: Freshwater Management in Aotearoa New Zealand.* London: Palgrave Macmillan.

Parsons, Meg, Johanna Nalau, Karen Fisher, and Cilla Brown. 2019. "Disrupting Path Dependency: Making Room for Indigenous Knowledge in River Management." *Global Environmental Change* 56: 95–113.

Porter, Libby, Sue Jackson, and Louise C. Johnson. 2017. *Planning in Indigenous Australia.* London: Routledge.

Taylor, Melvin. 1953. "No Maoris Need Apply – It's the Colour Bar. 1 September." *Auckland Star* 1953.

Thompson-Fawcett, Michelle, Jacinta Ruru, and Gail Tipa. 2017. "Indigenous Resource Management Plans: Transporting Non-Indigenous People into the Indigenous World." *Planning Practice & Research* 32 (3): 259–73.

Unknown Author. 1909a. "Native Land Problem." *Wairarapa Age*, May 5, 1909.

———. 1909b. "Orakei as a Suburb." *Auckland Star*, February 12, 1909.

———. 1912. "Orakei Suburb Proposal." *Auckland Star*, October 24, 1912.

———. 1924. "Model Garden Suburb." *New Zealand Herald*, December 27, 1924.

———. 1928. "Harbour Pollution." *Auckland Star*, March 1, 1928.

———. 1929. "Squandered Fortunes. 22 August." *Sun*, 1929.

———. 1931. "Orakei Village. 20 July." *Auckland Star*, 1931.

———. 1932. "Dangerous Oysters. 25 June." *Auckland Star*, 1932.

———. 1935. "Future of Orakei." *Auckland Star*, July 8, 1935.

———. 1937a. "Maori Housing Problems. 27 July." *Press*, 1937.

———. 1937b. "Orakei Housing Conditions." *New Zealand Herald*, April 24, 1937.

———. 1939. "Orakei Lands. Report of Royal Commission Appointed to Inquire into and Report as to Grievances Alleged by Maoris with Regard to Certain Lands at Orakei, in the City of Auckland." Appendix to the Journal of House of Representatives (AJHR), 1939 Session I, G06. New Zealand Parliament. Wellington: Government Printer.

———. 1940. "Typhoid Fever at Orakei." *Auckland Star*, July 16, 1940.

———. 1943. "Auckland Maoris Living in Squalor." *Northern Advocate*, January 27, 1943.

———. 1944a. "Orakei Maori Houses." *Gisborne Herald*, February 9, 1944.

———. 1944b. "Slum Houses." *Auckland Herald*, July 11, 1944.

———. 1950a. "Old Maori Pa at Auckland. 23 February 1950." *Ashburton Guardian*, February 23, 1950.

———. 1950b. "Orakei Pa Shacks Make Problem for Maori Department." *Gisborne Herald*, February 23, 1950.

———. 2017. "Disappointment Community 'shut out' of Large Housing Development on Auckland's North Shore. 28 November." *Stuff News*, November 28, 2017.

Veracini, L. 2010. *Settler Colonialism: A Theoretical Overview.* New York: Springer.

Viriaere, Hinetaakoha, and Caroline Miller. 2018. "Living Indigenous Heritage: Planning for Māori Food Gardens in Aotearoa/New Zealand." *Planning Practice & Research* 33 (4): 409–25.

Waitangi Tribunal. 1991. *Report of the Waitangi Tribunal on the Orakei Claim (Wai-9).* Wellington: Brooker & Friend Ltd.

Walker, Erana T., Priscilla M. Wehi, Nicola J. Nelson, Jacqueline R. Beggs, and Hēmi Whaanga. 2019. "Kaitiakitanga, Place and the Urban Restoration Agenda." *New Zealand Journal of Ecology* 43 (3): 1–8.

Waitangi Tribunal, 2010. *The Wairarapa ki Tararua Report.* Legislation Direct. Wellington.

Whata, J. 2020. *Ngāti Maru Trust v Ngāti Whātua Ōrākei Whaia Maia limited [2020] NZHC 2768.* High Court.

Wheen, N. R., & Hayward, J. (2012). Treaty of Waitangi settlements. Bridget Williams Books. Wellington.

Wolfe, Patrick. 2006. "Settler Colonialism and the Elimination of the Native." *Journal of Genocide Research* 8 (4): 387–409.

Yiftachel, Oren. 1998. "Planning and Social Control: Exploring the Dark Side." *Journal of Planning Literature* 12 (4): 395–406.

Suggestions for further study

- Independent Māori Statutory Board. 2021. "The Māori Plan." Independent Māori Statutory Board. https://www.imsb.maori.nz/what-we-do/the-maori-plan/.
- NZ History. 2016. "Obtaining Land." 2016. https://nzhistory.govt.nz/politics/ treaty/the-treaty-in-practice/obtaining-land.
- Robb, Andrew. 2018. "Bastion Point: A Desperate Struggle and a Dream Fulfilled." *E-Tangata* (blog). https://e-tangata.co.nz/history/bastion-point-a-desperate-struggle- and-a-dream-fulfilled/.
- Small, Robert Wayne. 2020. "Ngā Kite Hauora Nō Ngāti Whātua Ōrākei Pourewa Gardens." Unpublished Research Plan. https://www.researchbank.ac.nz/ handle/10652/5069.

11

COLONIZED PEOPLES

Planning and the informal Indian city

Sangeeta Banerji and D. Asher Ghertner

Introduction

Arthur Crawford, the first municipal commissioner of the colonial presidency town of Bombay, wrote in 1908:

> No sooner had we turned all dangerous trades out of the Island - except the common tanners, whom we located at Tannery Town (Dharavi) a very suitable position for the trade - No sooner had I proclaimed that no Factories of any kind should thence forth be built south of Elphinstone Road - No sooner. I say. had these essential steps been taken, then Kessowjee Naik brought his dyers back to their old quarters. I prosecuted them, but was defeated.
>
> (quoted in Dossal 2005)[1]

Crawford, a great advocate of comprehensive physical planning, here captured the obstinacy he experienced in his efforts to bring the spatially and politically riotous city to order in the aftermath of the anti-colonial Indian Rebellion of 1857. Each attempted modification of land-use patterns, he found, faced resistance from propertied classes, especially the owners of industrial units interspersed throughout residential neighborhoods. Crawford was eventually forced to resign from his position on corruption charges, hinting at the extreme lengths to which even colonial administrators had to go to "bring order" to urban India.

Principles of urban planning have long been imagined in India to be outside (and above) the rough and tumble of party politics, the "vernacular" economy, and the multiple, non-cadastral types of land uses common there. Yet, as in the 19th century colonial city, so in the contemporary neoliberal one, planning, has persistently been informalized, drawing planners into the vernacular

DOI: 10.4324/9781003157588-12

processes of rule interpretation and governance that make the distinction between planning and politics in post/colonial societies difficult to maintain. Be it the Improvement Trusts established in the early 20th century, the town planning organizations driving the developmentalist impulses of post-independence India, the numerous new towns built as the "temples of modernity" in the Nehruvian era of state-directed planning, or the rise of smart city urbanism today, the technical imperatives of planning have persistently succumbed to the pressures of class politics, generating new rationalizing schemes that only temporarily resolve to lift planning above the politics of city making.

This chapter draws on the rich history of planning in India to argue that informality, so often seen as a post-liberalization trend of privatized and splintered planning, has been a defining feature of planning the Indian city since at least the British colonial era. While viewing urban centers of India as the sites of failed urban planning is the norm (cf. Roy 2009), in reality an alternative planning discourse has silently operated that inserts the contentious demands of the urban majority into planning practice. This mode of alternative planning operates precisely via the vernacular mechanisms of rule-making and rule-breaking that bend zoning, property law, and building code to the needs of local circumstances that have historically been defined by global planning systems as planning's Other. Such informalized planning mechanisms are central to the Indian urban experience. Planning the (post)colonial city, in other words, often operates through the violation, even erasure, of the planning norms inscribed through European and U.S. town planning curricula and systems. This suggests not planning's irrelevance to postcolonial societies, but planning power as a central terrain on which democracy actively plays out (Sundaresan 2019).

The chapter emphasizes the persistent conflicts between the technical and the political that have characterized the Indian city. The first section describes how colonial urban planning principles were based on racial segregation, which were challenged by a rise of native self-determination in municipal councils articulated as a movement oppositional to the very idea of British-style planning itself. The second section discusses the origin and uptake of Improvement Trusts in colonial India, new British-led institutions that promised to use updated technical guidance to supersede the inequalities of the colonial city. The third section shows the persistence of a modernist technical vision for a secular India, but one that became heavily splintered, due to a rapid influx of refugees and rural migrants following the Partition of British India and uneven urban industrial development. The fourth and fifth section, discuss the most turbulent periods of modern urban India, which experienced a political emergency, economic liberalization and the growing influence of international financial agencies in urban policy. The chapter ends with current planning practices, which have seemingly abandoned the quest for social justice in urban India by asserting planning, yet again, as a techno-managerial regime – mirroring developments in Great Britain. Envisioning and creating more just cities requires attention not just to a range of rich experiments in inclusive housing and pro-poor land economies, as numerous Indian urban studies have productively shown (Benjamin 2008; Bhan 2019; Ghertner 2020; Mukhija 2001), but also attention to how planning's "failures" are and have been appropriated by the urban majority.

The late 1800s–1900: colonizers and the colonized in the Indian subcontinent

Though the Indian subcontinent has had a long history of urbanization dating back to 37 B.C., British Rule introduced numerous new urban forms. While Indian cities historically had been spatially structured along lines of caste, religion, and class (Dossal 2010; Kumar, Vidyarthi, and Prakash 2020), British spatial planning was organized primarily according to logics of economic extraction. After acquiring the *Diwani* rights, or the power to collect land revenue directly from the people in 1776, Lord Warren Hastings and Robert Clive, representing the East India Company, went on to declare Calcutta as the colonial capital. This marked the beginning of an era of resource extraction and colonial expansion in India, the effects of which are felt to this day. To expedite economic extraction and establish "dominance of dependence" in an already fractured landscape, the British created a regional hierarchy of urban centers to organize economic functions and extend administrative oversight.

The port cities of Mumbai, Madras, Surat, and Karachi rose in prominence as presidency towns soon after. Within those, segregated enclaves of white towns were established to house colonial officers, with native settlements designated as "black towns" to remain in a traditional, mixed-use, and dense pattern. White towns replicated the planned townships of Europe and benefited from premium infrastructure for water, sanitation, and green space. Native quarters, in contrast, were relegated spatially and symbolically to the sphere of customary practice, part of a racializing logic that deemed native settlements a mere extension of the primitivism assumed to characterize the village. The concentration of the colonial economy in these towns led to rapid population growth but minimal infrastructure investment. Consequently, sanitary conditions became so deplorable that by the 1860s Bombay acquired a global reputation as the "cholera's nest" (Dossal 2005; Dobbin 1972; Spodek 2013). To maximize resource extraction, the British also created an extensive urban and interurban railway network, which led to the emergence of market, canal, railway and industrial towns, different kinds of entrepôts within a cascading core-periphery colonial economic space.

The Indian Rebellion of 1857, though short-lived, had significant impacts on the urban and administrative institutions of colonial India. The Rebellion led to the British Crown's formal assumption of power and, in 1861, the nomination of Indian nationals, principally wealthy landowners, into the previously all-white legislative council. The passage of the Municipal Act of 1865 theoretically handed financial control of municipal revenue to these representatives. The Colonial Presidencies of Bengal, Bombay, and Madras were granted the power to appoint municipal commissioners to direct the financial resources stemming from local taxes toward urban expansion. While this marked an important extension of planning power to local representative bodies, earlier segregationist patterns were further ensconced into late colonial urban form via two trends.

First, due in part to accelerating urbanization and racialized fear of tropical disease in native settlements, military cantonments were set up for British soldiers, with civil lines established as officers' quarters (Legg 2008) – the latter tucked away from, but within proximity to, the native town. Designed to be visually and spatially differentiated from native quarters, the significant outlay on these colonial spaces also left municipal commissioners with limited resources to fundamentally reorganize urban form for non-Europeans.

Second, alongside the consolidation of urban industrial power, industrial elites and an upper-caste native gentry used their foothold in municipal government to build urban enclaves that enjoyed designated infrastructural services not afforded to the older urban centers or the expanding quarters of the industrial working class. Physical planning emerged by the late 19th century as a two-tier system of infrastructural haves and have-nots.

Alternate planning visions, articulated through an emergent working-class movement, aimed at redressing colonial and class segregation and became fused with broader national calls for self-determination and representative government. It is in this context that more democratic municipal institutions came to be seen as a foundation upon which an independent Indian state might emerge (Mandlik cited in Dobbin 1972, 135). Fierce contestations among local political, business, and caste associations broke out, however, over the limited colonial revenue available.

In the largest colonial presidency of Bombay, its first municipal commissioner, Sir Arthur Crawford, mentioned above, initiated various infrastructural improvements that reached only the wealthiest neighborhoods, despite drawing on increased taxation from the general population. The working classes were left in deplorable conditions that ultimately led to the outbreak of the Bubonic plague in Bombay in 1896. This catastrophe led to a retrenchment of colonial authority, with colonial governors seizing city planning powers back from municipal councils and to be placed in the hands of appointed Improvement Trusts aimed at building a bacteriological city using models deemed singularly European in origin. Democracy was not up to the demands of planning – not yet.

1900–1945: hegemonic urban improvement

Colonial urban planning in the first half of the 20th century was driven by the three principles of sanitary improvement, decongestion and nuisance removal. The localist ideas originating from orthodox contagionist doctrines, in which filth was considered the source of the plague (Kidambi 2004), drove an administrative effort to impose sanitary order over the city's burgeoning slums, the same areas systematically denied access to colonial infrastructural networks over the previous 50 years. The loosely framed building by-laws of the earlier municipal corporations, over which native representatives had gained some amount of control, were suddenly deemed inadequate to address the accelerating urban health crisis (Hazareesingh 2001). Improvement Trusts were posited as the remedy.

Centralized planning imperatives, most visible in the Improvement Trusts that emerged across colonial India in the first decades of the 20th century, drove the heavy interventionism of New Delhi, which emerged as the new imperial capital in 1911, when King George V employed the famous British architects Sir Edwin Lutyens and Herbert Baker to establish India's first master plan. The plan would extend colonial reason out across the existing city through a network of radial spokes knitted across 85 square kilometers of territory. The Madras Improvement Trust, established in 1920, aimed to decongest the southern coastal city by expanding into the suburbs anchored by more spacious colonial bungalows (Lewandowski 1979). The famous naturalist planner Patrick Geddes was employed by the Governor of Madras to achieve this aim. Geddes's unique approach directly informed the Bombay Improvement Trust's move toward "harmonious planning" (Dossal 2005, 168), which divided the city into "natural areas" intended to protect the landed, industrial elite from the working class through new suburbs such as Mahim, Dadar, and Parel. Such patterns, what Geddes would call an "inseparably interwoven structure," were premised on a vision of equilibrium, of balanced growth and mutually supporting trades and sectors. In this period, the logic of segregation shifted from that of race – dividing the colonizer from the colonized – to that of class (Kidambi 2001), pictured through a naturalist lens of preservation of existing functional zones, most already dominated by caste and class elites.

Though Improvement Trusts had lofty goals of creating public housing with good light and ventilation, their first actions were almost always slum clearance. The absence of coordinated resettlement led to the proliferation of hastily erected residences on the outskirts of these Trusts' newly planned areas (Hazareesingh 2001). Booms in cotton and grain prices during World War I increased the wealth of mill owners in colonial Bombay and Calcutta, who were also the primary landlords in the city. With the subsequent rise in land and housing prices (Chhabria 2019), urban renewal schemes failed to launch. Middle-income Indian families made up of doctors, court workers, and the intelligentsia found themselves driven into the housing built for the poorer classes, who in turn found their way into slums. Fierce spatial competition for housing in industrializing India led to rising class-based tensions, culminating in Bombay's 1919 textile mill strikes. Similar patterns endured until Independence in 1947, with the Great Depression slashing commodity prices, and the extractive relationship with the British Raj diverting colonial revenue further away from urban improvement efforts.

1945–1965: planning for independent India

The end of World War II and Indian Independence in 1947 coincided with the traumatic events of Partition. The Partition, enacted on religious lines to carve out a Muslim Pakistan from a secular-but-Hindu-majority India, forced millions of Hindu refugees from the newly formed Pakistan to seek shelter in Indian cities, generating an urgent need for public housing. Recognizing the deplorable living conditions in the refugee camps and an impoverished countryside left depleted by over two centuries of resource extraction, Indian leaders chose to adopt a consolidated development approach for reconstruction. Embodying Mahatma Gandhi's vision for rural development via cottage industries (Chatterjee 1993) and India's first Prime Minister Jawaharlal Nehru's vision of a future industrial "dreamworld," the project of national reconstruction was based on planned economic expansion. This involved a two-pronged approach toward developing agriculture and modernization through industrialization (Chakravarty 1993). The Soviet style five-year plan – rather than British liberalism – became the operational instrument of choice. The First Five-Year Plan (1951–1956) established multiple rehabilitation projects for Partition refugees, with a primary focus on Delhi, Bombay, Ahmedabad, and Calcutta. The Plan also directed the state governments to ensure housing provision for public sector workers, the ranks of which grew during this period through state-directed industrialization efforts (Batra 2009). The national development model of the early Five-Year Plans included an explicit focus on "social justice" via land reform and secular governance, but entrenched business interests led to the relative neglect of the urban poor.

The G.D. Birla Committee, established by the central government in 1950 to assess the planning needs of Delhi, offered a scathing critique of the colonial Delhi Improvement Trust's failure to provide adequate housing over the previous decades (Sundaram 2012). In doing so, it paved the way for comprehensive master planning, which brought India into seeming alignment with international planning trend at the time and was to be carried out by a new urban local body, the Delhi Development Authority. Post-colonial planning in India thereafter came to be directed by a mix of leading industrialists (Birla was one), national political leaders, prominent international planners, and a growing field of Indians trained in planning (the first School of Town and Country Planning was established in 1955). Delhi's first master plan, promulgated in 1962, and planning more broadly at the time, recognized the unavoidability of mixed land-use planning, which, by then, had been recognized as a defining feature of Indian urbanism. Despite the insertion of several progressive provisions in the Delhi master plan – state control of property, heavy restrictions on private land speculation, and provisions for sheltering the poorest sections of society in state housing – the colonial image of the slum as the central problem confronting the Indian city became a defining feature of post-colonial Indian planning. The tension between planned entitlements for the working classes and the continued illegalization of its principal residential space – the slum – would become a defining contradiction of planning throughout India in the decades to follow.

The Second and the Third Five Year Plans, known as the Nehru Mahalanobis Plans (1956–1966) under the aegis of Nehru and the chairman of the planning commission, were Soviet-style plans for economic development that directed significant resources to the creation of large-scale, capital-intensive projects. Large hydroelectric dams and shipyards were placed in close proximity to industrial townships that were planned to provide not just housing for workers, but also a Nehruvian socialist goal of "entirely new kinds of places inhabited by new kinds of people who would directly participate in the grand project of building the nation (S. Roy 2007, 134). Unlike the "old cities" central to the nationalist freedom struggle, these new spaces were designed to allow state techno-managerialists to design a new India from scratch. While the colonial Improvement Trusts were retained in most old cities to address slum growth, 120 planned and industrial townships (Kumar, Vidyarthi, and Prakash 2020) were launched as engines of industrial advancement and as an explicit alternative during this period.

Aided by the federal funding and colonial instrument of eminent domain for land acquisition, originally introduced by the British in 1898, these new towns were planned by famous European and American architects and planners as alternatives to the politically fraught and congested metropolitan cities. The construction of the capital city of Chandigarh by the renowned modernist architect Le Corbusier was symbolic of Nehru's faith in technical planning as a way toward modernization (Prakash 2002). Devoid of the fraught urban politics that had come to plague the old city centers of India, these spaces were imagined architecturally as "neither English, nor French, nor American, but Indian" (ibid). Unlike the master planning processes underway in existing cities, which depended on local municipal bodies to generate revenue, these new towns were funded substantially by the central government.

The newly established development authorities around the metropolitan areas of Calcutta and Bombay, aided by international funders such as the Ford Foundation, commissioned influential international planners, including Albert Mayer, Otto Keonigsberger, and Herman Goetz, as well as local intellectuals, to begin developing the new satellite townships of Navi Mumbai outside of Bombay and Salt Lake outside of Calcutta. These plans aimed to rationalize urban expansion in the nation's two biggest cities, casting them as alternatives to the urban concentration and social mix that dominated city planning previously. However, heavy expenditures on this whole suite of modernist projects – new towns, satellite towns, and industrial districts – meant that the spaces that the urban poor had come to inhabit grew rapidly in density (Chhabria 2019). While the national planning consensus advanced the dream of the socialist city (e.g., factory towns centering on the industrial worker), they simultaneously recommended that the urban poor, presumed to have "un-urban habits," be moved outside of the city limits (Sundaram 2012). Increasingly, in the decades following Independence, the legitimate housing for the poor was illegalized and demarcated as the modern slum (Björkman 2014; Weinstein 2014).

Steel towns were built by Hindustan Steelworks Construction Limited (HSCL), a special purpose vehicle established by the central government to construct and manage steel plants and associated civic infrastructures in what are today the eastern states of West Bengal, Jharkhand, Orissa, and Chhattisgarh (S. Roy 2007). Yet again, the workers, or "producer patriots," that came to inhabit these cities were reduced to economic agents, with differentiated land-use zoning deemed a solution to urban conflict: each space had its designated use and designated user. Technical parameters such as number of households required for the factories, distance from the sources of the raw material, and labor hierarchies defined how these cities were constructed and governed. However, instead of doing away with caste-, race-, and community-based settlements of the old cities, these new towns created new class-based settlements based on hierarchies of employment that were more insular that had been imagined (Sivaramakrishnan 1976).

By 1971, these townships had been declared largely as failures by national planning organizations due to the rising incidences of crime and political strife. This was the very object these townships' planners considered to have been "solved" by urban form and economic rationalization – much like the aspiration of the British colonizers to order the colonial city. The problem of communalism that Nehru imagined these planned cities would resolve would continue to shape his urban "dreamworlds" in the years to come.

1965–1980: Garibi hatao (eliminating poverty)

From 1965 to 1980, the megacities of Delhi, Bombay, Calcutta, and Madras saw an incredible population growth of 80 million, despite concerted efforts to divert in-migration (Batra 2009). The Fourth (1969–1974) and the Fifth Five Year Plans (1974–1979) focused on the creation of small towns to ensure a balanced spatial distribution of economic activity across the country, and a shift away from mass slum removal in favor of a "reconditioning of the slum." The New Okhla Industrial Development Authority (NOIDA) was designed as a satellite city on New Delhi's outskirts at this time. Simultaneously, the City and Industry Development Corporation (CIDCO) in Maharashtra was formed as a vehicle to deindustrialize and decongest the metropolitan core in the 1970s. Both NOIDA and CIDCO aimed to check land speculation and redirect migration by creating serviced land for planned neighborhoods. Unlike the master plans of the metropolitan cities they neighbored, CIDCO and NOIDA received direct federal funding, which made the process of land acquisition, pooling and servicing much faster. This model of parastatal agencies would later emerge as a prominent institutional fix to the "problem" of democratic politics intruding on the imperatives of physical planning.

However, the more technocratic visions of urban planning represented by CIDCO and NOIDA would have to wait, as a range of political upheavals placed the terms of urban democracy themselves front and center. The devastating Sino-Indian War in 1962, the death of Nehru in 1964, the Indo-Pakistani War of 1965, and droughts in 1966 and 1967 resulted in food shortages and a balance of payments crisis. This meant that the modernist project of urban development had to be postponed, as the newly elected Prime Minister Indira Gandhi renegotiated the statist and socialistic planning model embodied in the Five-Year Plans up to that point (Chakravarty 1993).

Elected to parliament with a landslide majority in 1971 behind the slogan of *Garibi Hatao*, or eliminating poverty, Gandhi promised to redistribute wealth via rural land reform and urban public housing expansion (Frankel 1978). Her ruthless efforts to consolidate political power, however, led to political unrest in the form of student agitations in the country's major cities. This situation, coupled with the 1973 oil crisis and another war with Pakistan in 1971, saw the imposition of an internal emergency in 1975–1977, in which civil rights were suspended. Low economic growth and the increasing dominance of what Kaviraj (1988) has called the "bourgeois state" – a coalition between the capitalist class and the bureaucratic intellectual elite – whose ideology was similar to that of the British colonizers, left the impoverished populations in rural areas with no choice other than to migrate to cities, despite state efforts to curtail this movement.

To address the growing inequality and control the spiraling land prices in urban India, the Urban Land Ceiling and Rent Control Act (ULCRA) was promulgated in 1976 to prevent land ownership concentration. Initiated during the emergency, ULCRA imposed a ceiling on the amount of vacant urban land individuals could hold.[2] Fashioned after the post-colonial land reforms meant to abolish the *zamindari* system of landlord power in rural India, ULCRA prescribed a limit of 500–2,000 square meters of land per individual. Any land above this limit was to be handed over to the state for token compensation or to be used for low-income housing. However, due to myriad exceptions and the generally poor state of land records in India, there were many hindrances in its implementation. Only 12,000 hectares were ultimately acquired of a recorded 166,162 hectares deemed eligible across the country, and only about 650 hectares were finally used to provide public housing for the poor (Siddiqi 2013; Srinivas 1991). Opportunities for land speculation thus emerged, which saw land prices rise even further.

During the Emergency, massive demolition drives displaced hundreds of thousands of people to the peripheries of India's megacities. In Delhi alone, approximately 700,000 people were displaced from their central-city neighborhoods, with resettlement for many contingent on agreeing to sterilization (Tarlo 2003). Similar demolitions displaced 20,000 people in Mumbai, but with different, more covert tactics. To avoid any dissent, the municipal corporation would demolish only one neighborhood at a time and shift the population to un-serviced plots on the outskirts of the city (Singh and Das 1995). In Bombay planners preparing updated land-use plans in 1967 demarcated large swaths of land as either green spaces or proposed roads that needed to be acquired. This act of mainstream planning has the effect of rendering invisble the populations that lived in these areas, setting the stage for mass demolitions that would follow when mechanisms of municipal democracy, many resting on close ethico-political ties between squatter communities and local bureaucrats and officials, were weakened or suspended.

The 1980s saw a marked transition in how the urban poor were figured in the urban imaginary, with the judiciary offering a corrective to both the overreach of the Emergency period and the gradual privileging of propertied classes over the urban poor. Marking the early phase of what Bhan (2016) calls "the juridical city," in which the judiciary would become a key interpreter of planning law and enforcer of planning practice, the Supreme Court of India issued a favorable decision in a public interest litigation filed by Olga Tellis, a journalist and housing activist, to stop the removal of pavement dwellers in Bombay. This period also saw the beginnings of interventions by the World Bank and the Asian Development Bank in major urban infrastructure projects, including the Madras Urban Development Project in 1985, designed as an alternative to slum clearance. In this project, renewable, 30-year leases were provided to slum co-operatives along with subsidized loans intended to support infrastructural upgrades (Bardhan et al. 2014). While this program was only available to slums located on public lands, sites-and-services schemes funded by the central government in other cities, most prominently Indore, involved the relocation of slum settlements to the city's outskirts on pre-serviced plots.

1980–2000: economic liberalization and planning reform

The failure of organized labor movements in urban India's industrial centers in the 1980s paved the way for a socio-spatial transformation of industrial land into high-end residential and commercial real estate in cities including Bangalore and Mumbai. Owing to the growth of economic sectors such as I.T. and financial services, the structure of the labor force in these cities changed from being overwhelmingly blue collar to white collar much like the United Kingdom under the neoliberal leadership of Margaret Thatcher. The mill lands at the center of most Indian megacities, which had originally been leased to industrialists by the colonial governments at diminutive rates, were sold as private property, defrauding the mill workers' living and working on that land for decades. The socio-spatial changes in cities in this period, driven by the technology industry's growth, signaled a seismic change in urban planning practices in India (Adarkar and Phatak 2005; Anjaria 2006).

These changes in city planning were spurred by the liberalization of the Indian economy in 1991 and the devaluation of the currency, yet again propelled by a balance of payments crisis. Moving away from the state-led provision of housing and infrastructure, in what could be understood as the end of the socialist development policy of India, market actors and international financial institutions, such as the World Bank and Asian Development Bank, started taking an active role in formulating urban policy. The Eighth Five Year Plan (1992–1997) introduced a system of "cost recovery" into municipal budgeting (Baindur and Kamath 2009), aiming to limit the funds received from the central government for infrastructure. By 2004, municipal governments were required to raise all funds for infrastructure development from market actors in bonds, justified as necessary to increase service-delivery and cost "efficiency."

In addition to promoting public–private partnerships for infrastructure provision, international actors pushed for amendments to the Land Acquisition Act, ULCRA, and the Transfer of Property and Rent Control Act. This period also saw the exponential growth of parastatal institutions, such as the Bangalore and Mumbai Metropolitan and Regional Development Authority (BMRDA and MMRDA) and special purpose vehicles that, unlike the colonial-era Improvement Trusts, operated on market principles. Free from democratic accountability and headed by high-level bureaucrats appointed to ensure "efficiency," these institutions ushered urban India into an age of neoliberal globalization.[3]

The Ninth Five Year Plan (1997–2002) put the onus of urban development squarely on municipal bodies' shoulders. In terms of land-use planning, despite the very low implementation rates of the master plans in metropolitan cities, there was a push to introduce market-based tools such as the Floor Space Index and Transferable Development Rights, popular urban policies in the United States, to bring financing to land acquisition processes (Pethe et al. 2014). Coupled with the real estate boom influenced by the gradual liberalization of the land market to foreign direct investment in the early 2000s, these provisions made the Municipal Corporation of Mumbai (BMC) the country's richest urban local body. The BMC nonetheless continued to advance a privatization agenda, specifically in the domain of water provision, placing extreme pressure on basic service access in slums. Guided by national policies that encouraged the private sector's involvement in the provision of housing, city governments initiated slum rehabilitation policies that would capitalize on the real estate boom, advancing the goals of land monetization underlying the policy directives from the central government (Batra 2009). In Mumbai, the Slum Redevelopment Authority was formed to re-house residents being displaced for urban redevelopment who could show proof of residence in settlements before 1995, but multi-story resettlement flats proved expensive and undesirable for many would-be resettles (Nijman 2008; Weinstein 2014).

Finally, a coalition of high-level bureaucrats, top politicians, and leading industrialists – acting similarly to the Indian elite who dominated the municipal councils in colonial India – were instrumental in advocating for a "world-class" urban future. For example, the "Vision Mumbai" proposal, published by the international consulting firm McKinsey and Company in 2001, articulated the specific goal of converting Mumbai into Shanghai, with the latter serving as a symbol of how to make beautiful and efficient a once-unruly city. This proposal was produced "at the request of Bombay First," an elite civil society organization representing business and property interests, "in concurrence with the state government, [and] recommended the uptake of world-class infrastructure to increase the standard of living in the city." These recommendations were followed by implementing large infrastructure projects that ultimately resulted in the displacement of approximately 300,000 low-income residents (Doshi 2013; Björkman 2015). Comparable initiatives were launched in Bangalore by the civil society group Janaagraha and the Bangalore Action Task Force. Similarly, in Delhi, a shift occurred in the "ethico-political" commitments of local state actors, from that of meeting the demands of the poor to appeasing the whims of local resident welfare associations, local property-owning groups increasingly antagonistic to the presence of slums (Ghertner 2011). These associations and the bourgeois state that backed them drove a renewed colonial sense of "nuisance law" in the early 2000s, which rationalized the displacement of as many as a million slum dwellers from Delhi on the basis of their aesthetic obstruction to Delhi's world-class aspirations (Ghertner 2015). While participatory institutions were being introduced within this period, there was an active disavowal of the space that the urban poor had come to occupy in the city (Coelho, Kamath, and Vijaybaskar 2011).

2010–present: privately planning the Indian city

The Jawaharlal Nehru National Urban Renewal Mission (JNNURM) was launched in 2005 by the Central Government of India, continuing its agenda of supporting the growth of small and medium towns. This program selected sixty-three cities with over a million people to receive upgraded infrastructural services (Batra 2009). By far the most expansive urban development agenda launched in post-colonial India, this program required the selected cities to pre- pare City Development Plans (CDPs) covering 20-year periods. These CDPs formed the basis for agreements between urban local bodies, the state government, and the central government. JNNURM provided seed money to help leverage additional funding from private investors. The intent behind the CDPs was to hold the local state accountable to a privately backed urban vision centered around land monetization and large-scale infrastructure development.

Core to this scheme was the requirement that participating state governments reform rent control and land ceiling laws and substantially reduce the taxes associated with the sale of land – each considered a legal-bureaucratic hindrance to a fully charged property market. JNNURM required that urban local bodies implement e-governance and property record digitalization. These demands were couched in the language of anti-corruption and accountability but were largely oriented toward particular market "efficiency" metrics, including the goal of increasing the rate of property tax collection, improving the speed with which land acquisitions could take place, and introducing user charges for basic services to fund targeted service expansion to the urban poor.

JNNURM was widely seen as the alternative to the centralized city planning instruments long in place. But arguably, as was the case with the municipal councilors in colonial India, JNNURM placed complex financing requirements and service provision mandates on urban local bodies while leaving large-scale decision-making in the hands of the corporate-bureaucratic elite. The existing master plans and development plans, which had achieved minimal implementation rates,[4] were seen as plagued by the encumbrances of "vote bank politics," from which this scheme would extricate the Indian city. Market efficiency and bureaucratic streamlining were neatly aligned, on paper at least.

In 2005, the Indian Parliament passed the Special Economic Zones Act with four main objectives: export promotion, investment encouragement, employment generation, and creation of world-class infrastructure (Kumar, Vidyarthi, and Prakash 2020). Special Economic Zones (SEZ), on the surface, appeared to be similar to the industrial townships planned in the 1950s. However, the scale of land acquisition involved, and the ability of the majority of the land acquired by the state, ostensibly for "public purposes," to be converted into private real estate, revealed a mutation in the mechanisms of regional planning toward what Levien (2018) calls "the landlord state." SEZs could range in size from one to more than ten square kilometers, and substantial exemptions were provided for corporate interests in privileged industries, such as pharmaceuticals and I.T. (Jenkins et al. 2015). While the intent behind this tool was boosting export earning, SEZs became an opportunity for corporate profit via state-sponsored land grabs.

More recently, Prime Minister Narendra Modi launched his "100 Smart Cities Mission," approving $14 billion to fund 100 new smart cities and the rejuvenation via smart technologies of five-hundred additional smaller cities. These funds were geared toward the incorporation of information technologies in the planning processes, followed free-market logics, and were to be divvied out on a competitive basis. This fast-paced model of economic development has been pitched against the conventional planning methods prevalent in Indian cities and envisions planning as an algorithmic practice removed from the pressures of local interest groups (Datta 2015). Though the "smart city" paradigm in most parts of the world has meant the incorporation of technological improvements, it has a much more flexible meaning in India, investing in "best practices" drawn from infrastructure-rich and financialized settings elsewhere. Alas, despite the declaration of this massive initiative as the new urban future, it bears a resounding similarity to urban planning initiatives of the colonial and post-colonial past, which repeatedly tried to lift the technical powers of urban planning above the messiness of politics.

Conclusion

This chapter covers nearly two centuries of urban planning history in India. While urban planning in colonial India was based on principles of racial segregation, the Indian propertied and merchant classes were instrumental in appropriating planning mechanisms to extend the benefits of selective infrastructure provision to themselves. Despite the growing dominance of democratic forces on urban planning decisions in the late 1890s, the creation of the Improvement Trusts in the colonial presidency towns was a critical conjuncture during which the struggle between the political and the technical within urban planning became definitive. The numerous Improvement Trusts, influenced by international planners including Patrick Geddes and the regionalist and Garden City movements popular in the United States and the United Kingdom, were designed to improve the living and environmental conditions of the urban poor through decongestion, regionalization, and sanitary improvement. These ideas of physical planning and suburban expansion continued to gain legitimacy via the master plans that shaped independent India.

After Indian Independence in 1947, a nation depleted by British colonization started its reconstruction process by adopting a consolidated approach that would focus on both rural and urban industrial development. Over the next few decades, guided by the vision of Nehru and inspired by Soviet modernism less than British liberalism, urban centers in India saw the mass establishment of planning organizations and 200 new towns built as "industrial utopias." Yet again, this was an attempt to circumvent the increasingly contentious environments within the metropolitan cities via technical solutions, much like the colonial period. While visions of socialist and secular urban centers were projected in the new towns, slums in the inner cities were increasingly displaced without rehabilitation or left to languish for future planners to deal with. This culminated in the slum demolition drives that characterized the Emergency of the late 1970s, when democracy was put on hold, and the machinations of physical planners and political masterminds coupled forced displacement with extreme forms of social engineering (Tarlo 2003) unlike any seen during British colonialism. Urban land-reform legislation, such as the land ceiling act, sought to reign in elite appropriation of the public city, but city planning's tight imbrication with economic planning, represented by the Five Year Plans, ensured that propertied classes always had the upper hand, much like their counterparts in colonial cities under colonial rule.

The next two periods discussed in this chapter saw the impact of the liberalization of the Indian economy in the early 1990s. Though the initial changes had already begun in 1980 via provisions for the conversion of industrial land uses into residential and commercial purposes, the large, infrastructural developments propelled by this economic shift came only after 2000. This is when land monetization rose as a central policy aim, triggering exponential growth of real estate for the next decade and far more punitive policies toward the urban poor, whose livelihoods were further constrained by the dislocation of urban industry. Market-based policies, such as the Slum Rehabilitation Scheme and Transferable Development Rights, emerged as alternatives to the federal funding mechanism that had been the foundation of urban planning in the previous decades.

Most recently, urban planning in India has turned toward the "smart city" paradigm, which seeks to retrofit existing cities with high-end infrastructure and technology-driven governance oriented toward reducing red tape, improving service delivery and cost recovery, and more generally producing more investor-friendly, world-class cities. Though the devices and policies have changed, there remain striking similarities between historical and present-day visions of urban planning as a domain that floats above the din of India's contentious sphere of popular urban politics. Yet, the will of the urban majority, articulated through the messy and contentious, has always found ways to shape urban planning toward more democratic aims.

Alternative planning history and theory timeline

Period	Planning history and theory canon	Alternative timeline: colonized peoples [Indian subcontinent]
Late 1800s–1900	Birth of planning	*Colonizers and the colonized* Prevalence of racial segregation in planning principles, self-determination of native Indians within municipal councils in presidency towns
1900–1945	Formalization of planning	*Hegemonic urban improvement* Creation and failure of improvement trusts to improve sanitary conditions and decongest cities
1945–1965	Growth of planning	*Planning for independent India* Independence and Partition (1947), constitution of national town planning organization, alternative vision of the industrialist utopia
1965–1980	Midlife crisis of planning	*Garibi Hatao (eliminating poverty)* Internal emergency and suspension of civil liberties, massive demolition drives, Urban Land Ceiling Act of 1976
1980–2000	Maturation of planning	*Economic liberalization and planning reform* Slum upgradation and improvement, transformation of industrial to residential and commercial land-uses in city centers, liberalization of economy, soliciting public participation in planning
2010–present	New planning crisis	*Privately planning the Indian city* Jawaharlal Nehru National Urban Renewal, Special Economic Zones, Smart cities program

Case study: counteracting technocratic planning in Mumbai

In 2011, for the first time, the Bombay Municipal Corporation (BMC) commissioned the preparation of the Mumbai Development Plan revision to a French planning conglomerate called Group SCE instead of carrying it out in-house based on the usual technocratic approach. While this firm undertook a very corporatized planning model, employing a digital database, the planning bureaucracy mandated public consultations at every stage of the revision process. While these mandates had merely intended to give lip service to participatory planning, a vibrant social movement gathered steam to challenge the principles governing the city and its planning thus far. When the existing land-use survey was first published, a coalition of civil society groups known as the Hamara Shehar Vikas Niyojan Abhiyan filed a record number of approximately 200,000 objections. Groups, as varied as elite civil society organizations, resident welfare associations, slum dwellers associations, informal workers, and tribal people, came together under one banner to highlight the survey's many mistakes (Kamath and Joseph 2015). While authorities tried to dismiss the objections because the members of the movement did not understand the basic principles of spatial planning, the public was able to frame explicit demands. The movement organized people around Mumbai's urban planning issues and socioeconomic well-being. It made the crucial contribution of training lay persons to read the technocratic guidelines of the development plan. For the first time, ordinary people could speak to planners about their lived reality. In 2015, the elected members of the bureaucracy rejected the plan prepared by the French consultants for the BMC. This outcome was a spectacular case of "bottom-up planning" in Mumbai, in which formerly invisible people took a small opening offered to them and spoke up to reframe the modernist principles of spatial planning for the postcolonial city.

Notes

1 This was Crawford's written response to a circular issued by Governor George Sydenham Clark's administration inviting suggestions from the Bombay Chamber of Commerce and the Municipal Corporation on the functioning of the Improvement Trust.
2 More than half the population of Bombay in 1989, for example, was living in slums, while 55% of the developable land remained vacant and in the hands of just 91 individuals (Srinivas 1991).
3 The landmark 74th Constitutional Amendment Act of 1992 was meant to carry out a true devolution of financial planning to the urban local bodies. This amendment was envisioned to elicit participation in urban policy by creating local area committees comprised of public-spirited individuals, not seasoned politicians. It turned out that parastatals essentially took over the "local governance" mandate of the 74th Amendment, albeit with none of the intended democratic wrapping.
4 In Mumbai, the 1992–1993 Sanctioned Plan had an implementation rate of just 12%–13%, despite having been a statutory law for over 50 years by then.

References

Adarkar, N, and VK Phatak. 2005. "Recycling Mill Land: Tumultuous Experience of Mumbai." *Economic and Political Weekly* 40 (51): 5365–68.

Anjaria, Jonathan Shapiro. 2006. "Street Hawkers and Public Space in Mumbai." *Economic and Political Weekly* 41 (21): 2140–46.

Baindur, Vinay, and Lalitha Kamath. 2009. "Reengineering Urban Infrastructure : How the World Bank and Asian Development Bank Shape Urban Infrastructure Finance and Governance in India." Bangalore.

Bardhan, Pranab, Michael Luca, Dilip Mookherjee, and Francisco Pino. 2014. "Evolution of Land Distribution in West Bengal 1967–2004: Role of Land Reform and Demographic Changes." *Journal of Development Economics* 110: 171–90.

Batra, Lalit. 2009. "A Review of Urbanisation and Urban Policy in Post-Independent India." Centre for the Study of Law and Governance, Jawaharlal Nehru University CSLG/WP/12 (12): 1–51.

Benjamin, Solomon. 2008. "Occupancy Urbanism: Radicalizing Politics and Economy Beyond Policy and Programs." *International Journal of Urban and Regional Research* 32 (3): 719–29.

Bhan, Gautam. 2016. *In the Public's Interest: Eviction, Citizenship, and Inequality in Contemporary Delhi.* Athens: University of Georgia Press.

———. 2019. Notes on southern urban practice. *Environment and Urbanization* 31(2): 639–54.

Björkman, Lisa. 2014. "Becoming a Slum: From Municipal Colony to Illegal Settlement in Liberalization-Era Mumbai." *International Journal of Urban and Regional Research* 38 (January): 36–59.

Björkman, Lisa. 2015. *Pipe Politics, Contested Waters: Embedded Infrastructures of Millennial Mumbai.* Durham and London: Duke University Press.

Chakravarty, Sukhamoy. 1993. *Development Planning: The Indian Experience.* New York and London: Oxford University Press.

Chatterjee, Partha. 1993. *Nationalist Thought and the Colonial World – A Derivative Discourse.* Zed Books.

Chhabria, Sheetal. 2019. *Making the Modern Slum: The Power of Capital in Colonial Bombay.* Seattle: University of Washington Press.

Coelho, Karen, Lalitha Kamath, and M Vijaybaskar. 2011. "Infrastructures of Consent: Interrogating Citizen Participation Mandates in Indian Urban Governance." *IDS Working Paper* 2011 (362).

Datta, Ayona. 2015. "A 100 Smart Cities, a 100 Utopias." *Dialogues in Human Geography* 5 (1): 49–53.

Dobbin, C E. 1972. *Urban Leadership in Western India: Politics and Communities in Bombay City, 1840–1885.* London: Oxford Historical Monographs.

Doshi, Sapana. 2013. "The Politics of the Evicted: Redevelopment, Subjectivity, and Difference in Mumbai's Slum Frontier." *Antipode* 45 (4): 844–65.

Dossal, Mariam. 2005. "A Master Plan for the City." *Economic & Political Weekly* 40 (36): 3897–900.

———. 2010. *Theatre of Conflict, City of Hope: Mumbai 1660 to Present Times.* New Delhi and London: Oxford University Press India.

Frankel, Francine R. 1978. *India's Political Economy, 1947–1977: The Gradual Revolution.* Princeton: Princeton University Press.

Ghertner, D. Asher. 2011. "Gentrifying the State, Gentrifying Participation: Elite Governance Programs in Delhi." *International Journal of Urban and Regional Research* 35 (3): 504–32.

———. 2015. *Rule by Aesthetics: World-Class City Making in Delhi.* New York and New Delhi: Oxford University Press.

———. 2020. "Lively Lands: The Spatial Reproduction Squeeze and the Failure of the Urban Imaginary." *International Journal of Urban and Regional Research* 44 (4): 561–81.

Hazareesingh, S. 2001. "Colonial Modernism and the Flawed Paradigms of Urban Renewal: Uneven Development in Bombay, 1900–25." *Urban History* 28 (2): 235–55.

Jenkins, Rob, Loraine Kennedy, Partha Mukhopadhyay, and Kanhu C. Pradhan. 2015. "Special Economic Zones in India: Interrogating the Nexus of Land, Development and Urbanization." *Environment and Urbanization ASIA* 6 (1): 1–17.

Kamath, Lalitha, and Marina Joseph. 2015. "How a Participatory Process Can Matter in Planning the City." *Economic and Political Weekly* 50 (39): 54–61.

Kaviraj, Sudipta. 1988. "Critique of the Passive Revolution." *Economic And Political Weekly* 23 (45): 2429–33.

Kidambi, Prashant. 2001. "Housing the Poor in a Colonial City: The Bombay Improvement Trust, 1898–1918." *Studies in History* 17 (1): 57–79.

———. 2004. "'An Infection of Locality': Plague, Pythogenesis and the Poor in Bombay, c. 1896–1905." *Urban History* 31 (2): 249–67.

Kumar, A., S. Vidyarthi, and P. Prakash. 2020. *City Planning in India, 1947–2017.* New York: Routledge.

Legg, Stephen. 2008. "Ambivalent Improvements: Biography, Biopolitics, and Colonial Delhi." *Environment and Planning A* 40 (1): 37–56.

Lewandowski, Susan. 1979. "Urban Planning in the Asia Port City: Madras, an Overview, 1920–1970." *South Asia: Journal of South Asia Studies* 2 (1–2): 30–45.

Levien, M. 2018. Dispossession Without Development: Land Grabs in Neoliberal India. *Modern South Asia.* New York: Oxford University Press.

Nijman, Jan. 2008. "Against the Odds: Slum Rehabilitation in Neoliberal Mumbai." *Cities* 25 (2): 73–85.

Pethe, Abhay, Ramakrishna Nallathiga, Sahil Gandhi, and Vaidehi Tandel. 2014. "Re-Thinking Urban Planning in India: Learning from the Wedge between the de Jure and de Facto Development in Mumbai." *Cities* 39: 120–32.

Prakash, Gyan. 2002. "The Urban Turn." *Sarai Reader 2002: The Cities of Everyday Life.* New Delhi: Centre for the study of Developing Societies.

Roy, Ananya. 2009. "Why India Cannot Plan Its Cities: Informality, Insurgence and the Idiom of Urbanization." *Planning Theory* 8 (1): 76–87.

Roy, S. 2007. Beyond Belief: India and the Politics of Postcolonial Nationalism. *Politics, History, and Culture.* Durham: Duke University Press.

Siddiqi, Faizan Jawed. 2013. *Governing Urban Land: The Political Economy of the Urban Land Ceiling and Regulation Act in Mumbai.* Boston: Massachusetts Institute of Technology.

Singh, Gurbir, and P. K. Das. 1995. "Building Castles in Air : Housing Scheme for Bombay 's Slum-Dwellers." *Economic and Political Weekly* 30 (40): 2477–81.

Sivaramakrishnan, K. C. 1976. *New Towns in India: A Report on a Study of Selected New Towns in the Eastern Region.* Vol. 13. Kolkata: Indian Institute of Management Calcutta.

Spodek, Howard. 2013. "City Planning in India under British Rule." *Economic and Political Weekly* 48 (4): 53–61.

Srinivas, Lakshmi. 1991. "Politics in India Working of Urban Land Ceiling Act, 1976." *Economic And Political Weekly* 26 (43): 2482–84.

Sundaram, Ravi. 2012. "The Slum as Archive: Revisiting the Social City of the 1950s." *Paper Presented at the 21st Century Indian Cities Conference*, University of California, Berkeley, 7.

Sundaresan, Jayaraj. 2019. "Urban Planning in Vernacular Governance: Land Use Planning and Violations in Bangalore, India." *Progress in Planning* 127: 1–23.

Tarlo, Emma. 2003. *Unsettling Memories: Narratives of the Emergency in Delhi.* Berkeley: University of California Press.

Weinstein, Liza. 2014. "One-Man Handled: Fragmented Power and Political Entrepreneurship in Globalizing Mumbai." *International Journal of Urban and Regional Research* 38: 14–35.

Suggestions for further study

- Baviskar, Amita. 2020. *Uncivil City: Ecology, Equity and the Commons.* New Delhi: Sage.
- Crowley, Thomas. 2020. *Fractured Forest, Quartzite City: A History of Delhi and Its Ridge.* New Delhi: Sage.
- Denis, Eric, and Marie-Helene Zérah (Eds.). 2017. *Subaltern Urbanisation in India: An Introduction to the Dynamics of Ordinary Towns.* New Delhi: Springer.
- Mehta, Suketu. 2009. *Maximum City: Bombay Lost and Found.* New York: Knopf.

12

COLONIZED PEOPLES

The struggle to reframe (neo)
colonial planning in Anglophone
Sub-Saharan Africa

Garth Myers and Francis Owusu

Introduction

The struggle for alternative planning in Anglophone Sub-Saharan Africa has been a struggle against a colonial mindset. Professional urban planning was introduced across the region during British colonial rule. After independence (which came for most British colonies in Africa in the late 1950s and 1960s), the struggle to assert an African urban identity for planning met head-first with economic crises in the public sector and global forces for privatization. Debt and state incapacity combined to undermine government-led urban planning efforts and foster the growth of informal urban development processes across the region's cities.

The 21st century has brought a return to the centrality of the state in urban planning alongside a return to blueprint master plans, but parallel to a considerable burst of investment from private capital and state-owned enterprises from outside of the previously dominant Western donor countries and former colonial powers. The inflow of foreign capital and the master plans are accompanied by new urban development ideas that are attempting to create urban utopias in and around many African cities. However, neither this infusion of foreign direct investment, much of it going into urban infrastructure projects and elite housing in those utopian communities, nor the latest round of ambitious state-led master plans have dissolved the prominence of unplanned informal settlements in the cities of urban Anglophone Sub-Saharan Africa. It is often within these unplanned areas where alternative planning resides, and what have been called "Afropolitan" visions of city life can be found (Nuttall and Mbembe 2008, 1).

This chapter examines the cities of Ghana, Kenya, and Tanzania as examples of urban planning dynamics in Anglophone Sub-Saharan Africa. The focus rests on the major cities – Accra, Nairobi, and Dar es Salaam – with

DOI: 10.4324/9781003157588-13

some examples drawn from secondary cities such as Kumasi, Mombasa, and Zanzibar. This designation itself, of Anglophone Sub-Saharan Africa, has colonial origins, as it marks out countries where English is the primary language of planning law and practice – and these are the former British colonies of Africa South of the Sahara. There was tremendous variation to the character of British colonialism, and there remains a substantial diversity to the urban dynamics for the roster of cities in these three countries, let alone in those of the rest of Anglophone Sub-Saharan Africa (Botswana, Gambia, Malawi, Nigeria, Sierra Leone, Somaliland, South Africa, South Sudan, Sudan, Uganda, Zambia, and Zimbabwe). South Africa was rather unique, with its longer and vastly more consequential period of white settlement, its combination of both Dutch and British colonies (eventually all brought together in 1901), its earlier (but racially circumscribed) independence in 1910, and its extreme repression under white minority rule in the apartheid era (1948–1994). The case studies herein exhibit a range that typifies the region's dynamics outside of this South African exceptionalism.

Late 1800s–1914: the establishment of colonial urban control

During this era, Sub-Saharan Africa experienced the Scramble for Africa, where European powers sought to extend their ephemeral coastal spheres of influence into the interior of the continent, in competition with rivals (Coquery-Vidrovitch 1991). For the British, this often meant reliance on concessionary companies such as the British East Africa Company, but by the onset of World War I, in most cases, company control had given way to foreign office or colonial office oversight. Very little in the way of formal, professional urban planning took place, but at the same time, there were examples of the establishment of new cities for administrative purposes (Myers 2003). One major consequence in urban design was that colonialism laid the foundations for many African cities becoming dual cities, especially where European settlement was widespread, with European and African urbanisms side-by-side.

Typically, this dual city had one side dominated by permanent structures and a regularized plan, whether a gridded or more curvilinear street plan following British or French (in a few cases, Portuguese, Belgian, Italian, or German) ideas of modernist late-19th century urban order. The other side would have a largely irregular form, often devoid of wide streets at all, with an entirely vernacular built environment dominated by impermanent or semi-permanent structures. Morton (2019) analyzes, for example, the dramatic contrast in Maputo between its wealthier, European-oriented City of Cement and the poor, unplanned, African-dominated and ultimately much larger City of Reeds: the place-names of the two sides say it all. Other cities showcased a more modest imprint of European urban form adjacent to or atop pre-colonial African ones (O'Connor 1983). This laid the groundwork for the persistence of unplanned urban space that would eventually come to be called informal settlements (Myers 2011).

In our countries of focus, the colonial era exhibits the diversity of experiences. British influence in Zanzibar gradually overtook American, German, and French imperial ambitions and established a Protectorate over the Omani Sultanate of Zanzibar in 1890, but only an 1892 sketch of the Arab, Indian and European side of town, erroneously labeled as a "plan," exists as a legacy of formal planning in the era (Bissell 2011). Mainland Tanzania became the colony of German East Africa and remained as such until the end of World War I. The Germans made the trading town of Bagamoyo their initial capital, only shifting to the moribund 19th-century Omani-Zanzibari colonial port of Dar es Salaam in the late 1890s but not investing substantially in planning for it (Brennan and Burton 2007). British East Africa (which would not become Kenya Colony until 1920) had its initial capital in the thousand year-old port of Mombasa until the colonial office shifted it to a new setting halfway up the new Uganda Railway in a swampy plain, Nairobi, in 1901. However, the new capital had no formal plan until 1926 (Smith 2019).

In Ghana, Towns Ordinance was introduced by the British Administration in 1892, enabled the government to acquire land for public works, collect revenue and undertake health and sanitary measures in the towns. The Towns Council Ordinance was also passed two years later for larger urban agglomerations (Dickson 1969). The main focus of city planning in Ghana was Accra, a Ga town long before it became a colonial capital. Many Europeans, including the Dutch, the Portuguese, the French, the Swedish, the Danish, and the British, had forts there. However, in 1874 the British seized control of Accra and moved their colonial capital from Cape Coast there to protect Europeans from native-borne diseases. Based on the Public Lands Ordinance, British officials carved out space within the city for expatriate and elite African traders who embraced European modernity aesthetics. However, the colonial officials could not decongest the more crowded parts of the old town through planning, regulation, and political negotiation, until the 1894 fire. This fire allowed them to remake parts of the city based on European standards by ignoring indigenous cultures and practices of placemaking. The latter was portrayed as "one compact mass of thatched buildings arranged in a haphazard manner and separated by narrow crooked streets" (Stanley 1874, 77, cited in Grant and Yankson 2003).

In each of the case study settings, then, this early era was largely absent of comprehensive planning. As a result, a variety of urban design outcomes were manifest, but at root a repeated pattern existed wherein towns mainly grew according to customary practices, religious principles (including Christian, Muslim, and Indigenous religious principles), economic functions (typically relating to transformative long-distance and even global trading in the late-19th and early 20th centuries), and power dynamics at the local level (Cristofaro 2020; Myers 2016).

1918–1945: the formalization of colonial planning

The interwar years are often characterized as the period of the establishment of colonial control in British Africa, and this included in the realm of planning. Formal, professional planning arrived in the part of the region under study in this era. Henry Vaughn Lanchester (1923) developed a Study in Tropical Town Planning that was called the first formal plan for Zanzibar. Walton Jamison developed a similarly formal (and similarly meager) plan for Nairobi in 1926 (Smith 2019). Although neither can be described as a comprehensive master plan as understood today, these plans and similar ones for Mombasa and other cities in British East Africa set in motion the development of building rules, establishment of limits of alignment, and the creation of boards for management and planning. The "tropical" dimension of plans like Lanchester's principally consisted of an obsession with sanitation and health, chiefly translating as planning for means of protecting white colonial officers from tropical diseases. Alongside these early plans, colonial regimes experimented with small-scale planned neighborhoods or programs for housing construction (Cooper 1987).

The colonial regime's first serious efforts for structuring Ghanaian cities began after World War I with the launch of the Ten Year Development Plan (1920–1930) and the passage of the Town Planning Ordinance. However, since the colonial administration investment decisions were influenced by access to natural resources, the provision of necessary infrastructure was limited to regions with exploitable and exportable resources. Thus, although these plans broadened the country's planning efforts beyond Accra, they focused attention on the major seaport towns and other cities in southern Ghana, including Kumasi. The plans also anticipated building new towns along the Accra-Kumasi railway and the building of mining towns by mining companies based on strict buildings and settlement rules. However, the towns in the Northern part of the country were generally ignored, which laid the foundations for the marked disparities between cities in southern and northern Ghana (Dickson 1969).

Despite the establishment of some plans, ordinances, building codes, and other vestiges of formal planning the reality in most cities was one where most urban growth took place outside of the colonial state's control. Where colonial regimes had substantial economic interests, such as in port cities or mining centers, more formal urbanism emerged in the form of comfortable villas for white adminis-trators in leafy neighborhoods with regularized street-grids, sewerage, drainage, and (eventually) electricity, but in most cases in disparaging contrast with the conditions for the majorities. The Garden City movement had a significant influ-ence, for instance in the creation of Lusaka as the new colonial capital for nearby Northern Rhodesia (today's Zambia) as a "Garden City for Africa" (Bigon and Katz 2016; Myers 2003, 2016). British colonial regimes in Gold Coast Colony, Kenya Colony, Tanganyika Territory, and the Zanzibar Protectorate (as these colonies were named in this era) held to highly segregated visions of the city. The patterns of these plans typically entailed a simple and dense grid form for African areas with more radial or organic – and far less dense – forms in white areas. Small segments of cities were set aside, also in densely gridded areas, for the Arab or Asian intermediary population.

Even where there was little to no white settlement, urban space was carved into segmented zones based around race and ethnicity. African migration to urban areas was highly restricted, but in spite of these restrictions, African areas grew in cities and towns of all four colonies. The character of urban settlement for Africans – while scarred by deprivation in comparison to white or Asian/Arab zones – manifested indigenous ways of living and being largely outside the purview of colonial rule. For example, along the Swahili coast of Kenya and Tanzania (including the Zanzibar islands), the Muslim majority utilized Islamic principles such as the "avoidance of harm" and respect for privacy, alongside customary practices of neighborliness and interdependence, to establish the pattern of urban growth in the absence of formal planning implementation (Myers 2003).

1945–1965: planning and movements for independence

After World War II, British Africa experienced what has been termed the "second colonial occupation" (Low and Lonsdale 1976), a period of sustained investment by the British designed to win over the colonized people as they prepared colonies for independence. In the Cold War context, the British held to a practical political concern with Africans in the transition to self-rule: "to help them remain stable and friendly" as "our greatest bulwark against communism in Africa" (Cohen 1959, 114). This meant a significant outlay of expenditures for new ports, airports, schools, hospitals, highways, and other urban infrastructure, and the planning frameworks to match (Cooper 1987). This was the era in which one can most readily identify the three "ideologies" that Robert Home (1997, 3–4) argued were at the heart of British colonial urban planning around the world, prioritizing "state control," capitalism, and "utopian … forms of social organization." Planning set about to create an enframing order, to make cities "readable, like a book" of containerized and segmented zones, to separate who could and could not be in the city, and to produce maps and regimes of surveillance, all the while maintaining the primary objective of profit accumulation for the colonial power (Berman 1990; Mitchell 1988, 44; Myers 2003). In city after city, though, the enframing efforts of the second colonial occupation fell short, and cities were reframed – largely to reflect the will of the urban majority and its growing demands for political power (Myers 2003). It was the decisions of "thousands of individuals and families," rather than of professional planners, that shaped urban development (O'Connor 1983, 237).

The first comprehensive master plan for Dar es Salaam was approved in 1949, drawn up by London's Alexander Gibb and Partners, paid for by the colonial metropole, and "bestowing urban form and social welfare from above" (Armstrong 1987, 134). Gibb's firm produced a number of ambitious plans, such as the plan for the development of a new major port city for Tanganyika in Mtwara on the southern coast (Alexander 1983). Zanzibar Protectorate's Ten-Year Development Plan for 1945–1954 was primarily a de facto master plan for that city; a further master plan was approved in 1958. Nairobi's colonial rulers hired the Cape Town-based English architect L. W. Thornton White and a team to produce a plan for a settler "colonial capital" garden city (White et al. 1948). None of these grand plans achieved much of the enframing they were designed around. Gibb's plan for Dar was notably "limited" in implementation (Armstrong 1987), White's team in Nairobi led even the colonial settler apologist writer Elspeth Huxley (1948, 251) to remark that they had barely made "the best of a bad job," and Kenya soon found itself in an "Emergency," a de facto war for independence from 1952 to 1960. Both the postwar development plan and particularly the 1958 Master plan for Zanzibar ran aground in what Zanzibaris term the "Time of Politics" that brought increasingly violent contestation between political factions in the city (Glassman 2011).

In Ghana, the colonial administration strengthened the legal framework for planning with the 1945 Town and Country Planning Ordinance. This ordinance was based on British planning legislation and was designed to modernize the society by enforcing strict spatial order and providing infrastructure. However, there were efforts to separate native settlements from colonial residences and promote European standards of design in the latter. For instance, Accra's building codes in areas of the city inhabited by Europeans were largely based on European characteristics. European-inhabited neighborhoods in Accra such as Ridge and Cantonments were planned as low-density developments. In contrast, communities such as Nima were left crowded and mostly unregulated, creating a shanty-town landscape. Such segregation was also evident in most of the major towns where the colonial administrators worked and lived including, Kumasi, Cape Coast, Sekondi-Takoradi, and other cities (Adarkwa 2012; Fält 2016).

Ghana attained independence in 1957, led by Kwame Nkrumah, who pursued socialist policies and sought to transform the country through state-led industrialization. The desire to achieve industrialization led to efforts to change Ghanaian cities' colonial functions from commerce to industrial production. It also meant improvements in the connection between different parts of the country. For instance, Nkrumah's national plans for economic development led to the development of townships of Tema, which was built in connection to the construction of Tema harbor and Akosombo to house the workers of Akosombo dam. Unfortunately, the effort was still concentrated in the southern part of the country and therefore did not address the north-south disparity created during the colonial era. The first real master plan for Accra, developed in 1944 but not implemented before the country's independence, was updated by the newly independent administration in 1958 and titled "Accra: A plan for the Town" (Fält 2016).

Thus, no sooner had colonial regimes begun to attempt implementation of urban plans of the second colonial occupation when the very people whose hearts and minds they were meant to persuade that the British were "there to help" rejected the plans (Myers 2003). Movements for independence (coming first to Ghana in 1957, then to Tanganyika in 1961 and Kenya and Zanzibar in 1963) were often centered in urban areas, and leaders and activists sought to reframe or completely upend colonial plans. Yet, as the Ghanaian example for Accra above suggests, there was an often surprising degree of continuity in formal planning in the immediate aftermath of colonial rule. There were few local architects, planners and engineers, and invariably most had been trained in British institutions. Even in socialist Tanzania, an "insensitivity" to the perspectives of the urban majority predominated in urban planning (Banyikwa 1989).

1965–1980: planning in the early independence period

The date, 1965, is a fuzzy beginning year for a new era in Anglophone Africa, and especially so because all of the countries in focus in this chapter had gained independence by then. In all three cases, though, the early independence period evolved dramatically in the 1960s as the decade moved on. Kenya gained independence in 1963 but 1969 marked a clear transition toward de facto single party rule. Zanzibar also gained independence in 1963 but had a bloody revolution a month later (January 1964) and united with Tanganyika three months after that (April 1964) to form the United Republic of Tanzania. In 1967, that United Republic embarked on an ambitious new development plan based on principles of an African socialist ideology labeled, in Swahili, *ujamaa* (family-ness). Under the leadership of President Julius K. Nyerere, Tanzania sought to create an alternative African political order and development path beholden neither to the West nor to the Eastern bloc. A centerpiece of ujamaa following the Arusha Declaration that launched it was the construction of a new capital city for Tanzania at Dodoma, in the middle of the mainland (Myers 2011). Dodoma was designed as a low-key, non-monumental farming town, befitting the philosophy of self-reliance, collective work and interdependence at the heart of ujamaa.

Dar es Salaam, disregarded as it was in the ujamaa era (1967–1985), nevertheless had two Master Plans designed for it in these years – 1968 and 1979 – oddly, by North American planners who were nonetheless sympathetic with ujamaa's ideals (Armstrong 1987). Socialist planning also predominated, but in a more starkly East European style, in Zanzibar, where planners from the communist government of East Germany wrote the 1968 master plan (Myers 2003). Zanzibar hosted consular offices and had much closer relationships with the Soviet Union and other Eastern bloc countries.

Nairobi, by contrast, was now the capital of an independent Kenya with a staunchly capitalist orientation. It's much more vertical CBD came to be dominated by the skyscrapers of major global banks, hotel chains and corporations, along with the iconic Kenyatta International Conference Center, whose tower lies at the center of the Nairobi skyline. Designed by Kenyan engineers and commissioned by the country's first President, Jomo Kenyatta, KICC was the tallest building in Nairobi for nearly three decades (it is currently the sixth-largest, but still the most-recognizable skyscraper). Both the tower's roof and the adjoining amphitheater take their inspiration from a thoroughly modernist adaptation of the design features of an African rondavel or circular hut (Adjaye 2011). Planning for the city nevertheless came within a rubric of national development planning that was still strongly state-led (Obudho 1982, 1983).

While Ghana had been independent for nearly a decade, the character of its independent regime changed dramatically in 1966 due to series of coups and political instability. Despite these changes, the country's policies were generally driven by a state-led economic development agenda until the 1980s. The development plans of this period aspired to achieve economic development through industrialization and nation-wide provision of infrastructure and services. The government was a major provider of housing in the major cities through the development of housing estates and provision of subsidized loans.[1] As national development strategies gained more attention during this period, they largely de-emphasized planning in big cities like Accra.

Once again, the ambitious, top-down plans of authoritarian regimes of various ideological stripes met their match in the streets. In other words, most of urban Kenya, Tanzania, and Ghana worked under a different rhythm, outside of state plans. Nairobi was championed as a "self-help city" whose metropolitan urban form reflected ordinary people's capacity to develop their own spaces far more than it did the authoritarian planners and elites (Hake 1977). Dar es Salaam became one of the world's most rapidly growing major cities – at precisely the moment the United Republic of Tanzania turned it back on it. Zanzibar more than quadrupled in population, and the vast majority of new residents found plots and built on them outside of planning laws or building codes, whether socialist or colonial in origin. Accra's decline in living standards and the quality and availability of housing compelled many residents to defy planning regulations and to build on vacant and unauthorized land, reducing official land use planning efforts to the small state-controlled lands. The result was a sharp contrast between the planned high-income, low-density residential areas such as Ridge, Cantonments, Airport Residential Area, and Labone and the low-income high-density unplanned residential areas such as Nima, James Town, and Sabon Zongo (Grant and Yankson 2003).

On the whole, then, the early Independence era witnessed massive urbanization processes and an uptick in national-scale interest in development planning. However, the national-scale development agendas were unevenly attentive to urban planning. What urban planning that did happen failed to keep pace with the growth of cities, and the quality of life for the growing urban populations declined. Poverty, precarity, and informality became prevalent features of urban life.

1980–2000: urban planning amid structural adjustment

By 1980, governments across Anglophone Africa were under duress, as states failed to make payments on public debts incurred largely in efforts to overcome colonial legacies in development. World Bank and International Monetary Fund structural adjustment programs (SAPs) designed to curtail government spending, reduce government involvement in the economy, devalue currencies, increase exports, and emphasize agriculture had severe consequences in cities. National and local governments lost large realms of their autonomy to the international financial institutions and bilateral donors, and urban services (the provision of water, sanitation, transportation, electricity and solid waste management, for example) shrunk dramatically or were forced into schemes of privatization (Briggs and Yeboah 2001).

The early independence period's enthusiasm for grand master planning began to wane in the late 1970s and 1980s. Therefore, as African states felt the combined pressures of debt, structural adjustment, and economic decline, a marked disinvestment in urban planning – and in master plans in particular – came to the fore. Where master planning took place, the plans were rarely implemented, and in their place came smaller-scale sectoral plans and schemes for privatization. In tune with the transformation in planning around the world, many African cities abandoned blueprint master plans for more participatory and focused action plans that targeted specific sectors or neighborhoods and were built around a variety of partnerships between public, private, and popular sectors – in other words, a curious mix of activist and neo-liberal planning predominated (Myers 2015).

In Dar es Salaam, when the local government sought aid to update the 1979 master plan, they approached the United Nations and in effect had their plan dictated to them by United Nations (UN-Habitat). What emerged became, as of 1990, the pilot program for the UN Sustainable Cities Program, which also came to Zanzibar and Accra (Burra 2004; Dar es Salaam City Commission 1999; Myers 2005; Ngware and Lusugga Kironde 2000). In Zanzibar, the last-gap socialist plan produced by planners from the People's Republic of China and published in 1982 faced delay upon delay in implementation (Kequan 1982). Instead, planning priorities shifted toward sectoral planning for re-developing the whole of the two islands that comprise the polity – but especially its historic urban core, Stone Town – for international tourism as a way to open Zanzibar's economy under structural adjustment (McQuillan and Lanier 1984). In Nairobi, national development plans and local sectoral plans continued to be unwritten, in effect, as they were published as structural adjustment bit into the state's capacity and autonomy (Charton-Bigot 2010; Huchzermeyer 2011). Ghana began implementing SAPs in 1983. These austerity policies had a devastating influence on Ghanaian cities, including the deteriorating infrastructure. The liberalization policies made it easier for individuals to build houses and run businesses, leading to many developments across Accra. They have also fueled the on-going explosive growth of new suburbs as well as slums (Grant 2005).

The failures of planning in the SAP era were, arguably, greater than those at any other time in Anglophone Africa's urban planning history, and they coincided with huge rates of urbanization, to produce a massive expansion in informal settlements across the region. With that informality came the blossoming of alternative ideas of what cities mean and what they could be, more than ever before, but within constraints of poverty, inequality and absence of political power for the urban poor majorities in many cities. One might consider the resulting urban form of massive informal settlements like Kibera or Mathare in Nairobi, for example, to be examples of alternative planning. Kibera, in that regard, came to be organized into 14 separate segments called "villages," and each village often has a predominant association with one ethnic group among Kenya's many ethnicities. A few Kibera villages have been planned or re-planned, but others have an irregular, clustered form with a social structure linked to that dominant identity (Mutisya and Yarime 2011).

The deprivation, insecurity, disenfranchisement, and environmental crises attendant with these developments provide a major caveat to any endeavor to champion this alternative. Kibera, for instance, developed a high crime rate, with substantial levels of violence, sharp animosities between villages, and severely negative environmental and health conditions that ossified in this era (Médard 2010; Njeru 2010; Owuor and Mbatia 2012; Simone 2004). Cities in all three countries came to be plagued by low-level corruption, organized crime, and illicit service organizations, often intertwined with local ethnic group structures, further reducing the capacity for claiming local ways as constructive alternatives to planning.

2000–present: crises and opportunity in Africa's "renaissance" era

The 21st century has been a weird and unpredictable roller-coaster ride for urban Africa. Economic growth at the macro level has been substantial in Kenya, Tanzania, and Ghana, and a new roster of investors, mainly from China but also from Turkey, Russia, South Korea, South Africa, and more "expected" sources like Britain and the United States, have utterly transformed he surface of the major cities of all three countries. Major new government master plans have emerged alongside elaborate new private master-planned satellite cities or gated suburbs in Accra, Dar es Salaam, Nairobi, Zanzibar, and elsewhere. In the 21st century, the left-for-dead era of grand blueprint master planning has seemingly reemerged across the continent. Some of the enthusiasm for longer-term, broader-scale city-region-wide planning is a product of the cash availability amid globalization and the "new Scramble for Africa" that revolves around natural resource extraction (Carmody 2011).

As global powerhouse economic actors pay greater attention to the continent, planning for its cities in the interests of these outside powers and investors becomes more common, such as in the UN Habitat (2013) white paper, *Unleashing the*

Economic Potential of Agglomeration in African Cities. This UN report largely adds to the "unbridled optimism of reports by private sector think tanks" about Africa's cities as "the linchpin" for economic growth as long as "governments can get their acts together to invest in in the 'right' kinds of basic and connective infrastructure ... to make doing business easier" (Pieterse and Parnell 2014, 14).

Yet this imagined privatopia renaissance in urban Africa is misleading. Much of it is a fantasy-land of computer-renderings and wishful thinking. Instead, ordinary residents of these cities have been whacked by crisis upon crisis (Pieterse 2010; Watson 2014). Poverty, climate change-induced flooding, political violence, the COVID-19 pandemic, and more crisis points are the actual defining features for the marginalized majorities of the region. The grand schemes often displace more participatory, action-oriented, and sector-focused planning.

However, despite the never-ending wave of crises, there are opportunities for alternative planning that, if they have not stopped the wave, have at least enabled pockets of hope to emerge and be sustained. Even by the latter half of the 1990s, but then especially through the 2000s, radical forms of decentralized, activist planning (or anti-planning) gained ground in many cities, often outside the state, even while various forms of urban governance linked to neoliberalism held increasing sway on the continent.

Perhaps the most dramatic example for this contradictory era in East Africa resides in Nairobi. Nairobi has been growing rapidly, globalizing and democratizing even as its poor majority continues to struggle with basic needs and its political system suffers occasional seismic shocks, whether from terrorism or political party violence (GoDown Arts Centre and Kwani Trust 2009). Kenya now lives under a progressive 2010 constitution that, on paper, decentralized a considerable amount of authority to the county-level, including Nairobi City County (Myers 2015). Yet palpable doubts about the last several national elections and the disconnect between elite visions and ordinary residents have endured and grown. Nairobi's residents, "caught up between the remains of empire and global city fantasies," thus have had deep-seated ambivalence to the latest wave of grand-scale planning, and especially the grand master plan of 2008, *Nairobi metro 2030: A world class African metropolis* (Smith 2019, 3). As planner Peter Ngau (2013, 17) put it:

> planning is centre-stage in this new era. Given the history of planning in Kenya, it is important that planners now act with integrity. We must ask ourselves fundamental questions. Are we, as planners, going to champion the changes brought about by the new [2010] constitution or are we again going to serve the interests of the rich minority?

There were certainly dynamic, innovative, and progressive local plans set in motion in different parts of Nairobi in the 21st century (see Boxed Case Study 1). The formal government planning for the city, though, has provided a more doubtful answer to Ngau's question, particularly through the *Nairobi metro 2030* plan (Myers 2015). Gaps between the formal plan for Nairobi and the realities of everyday life were glaringly obvious. Despite the huge emphasis on transportation planning that it set in motion, *Nairobi metro 2030* made no mention of the massive, thriving *matatu* (mini-bus) transport sector that serves most of the city and particularly its informal areas (Huchzermeyer 2011). Perhaps the most intriguing lacuna has to do with informality. The phrase, informal settlement, occurs once, for a city with dozens of profoundly poor slums housing an estimated 40% of the population (Owuor and Mbatia 2012, 125). The plan offered these areas next to nothing, with the possible exception of eviction to implement transportation plans. As Huchzermeyer (2011, 236) put it, in *Nairobi metro 2030*, "a land-hungry and unsustainable modern spatial order continues to be rolled out. Those evicted to make way for this spatial order can stage only a weak challenge."

It is notable to compare the results from the *Nairobi metro 2030* plan's supposedly participatory planning process (Republic of Kenya 2008) and the GoDown Arts Centre's community-created multi-site festival, entitled *Nai Ni Who?* (Who is Nairobi?), in 2013. According to Joy Mboya (2014, 69), "the festival was all about allowing people into processes and making space to provide input." She contrasted its popular energy and engagement with the county's master plan, which largely failed to generate popular participation, public comment, or attendance at open events. What could Nairobi be if its elites were not so caught up in achieving global status, and instead were more engaged in the dynamic participatory planning agenda of youth organizations, women's organizations, the GoDown Arts Centre, and other development NGOs in the city? There is a considerable degree of readiness, primed by political activism, for poor, informal settlement residents to engage in transformative, democratized planning at the grassroots, but these grassroots are disconnected and estranged from much of the formal state-led planning, left out of the vision for Nairobi in 2030. Similar contradictions can be found in Mombasa and other urban areas of Kenya, and they are present in a comparable form in 21st century Dar es Salaam, Zanzibar, Dodoma, and other urban areas of Tanzania.

The case of Old Fadama in Accra illustrates how informal communities can use civil initiatives to get planning officials to dialogue with them in addressing their needs. An informal community mostly for migrants of different ethnic groups from Northern Ghana, Old Fadama has been described as "one of the city's most deprived but innovative slums" (Nunbogu, Korah, Cobbinah, and Poku-Boansi 2018, 35). The community, which includes over 80,000 people located on public land, was considered illegal by the Accra Metropolitan Assembly. As a result, the residents lived with the constant threat of demolition and were deprived of state-supplied social facilities. The community's efforts in addressing its vulnerabilities started in 1993 when the members began to organize themselves into savings groups to promote social cohesion among the women and between the various ethnic groups in the community. The initiative gained the men's support, and by 1998, the community had developed a strong social and political hierarchy and began to secure official government recognition. The community elected its leaders, hired an office space, and in June 2012, registered the community under the Attorney's General Department as "Slum Union of Ghana." The union's primary functions include lobbying with the government to end forced evictions, rapid response to emergencies such as fire outbreaks and floods in slums, educating and sensitizing slum dwellers on their rights and responsibilities, and attracting social amenities to slum communities. The Old Fadama leadership successfully allied with other internal slum organizations, including the Centre for Public Interest Law, Shack/Slum Dwellers International, and People's Dialogue on Human Settlements. They effectively resisted Accra Metropolitan Assembly's eviction efforts, developed a cordial relationship with AMA, and implemented several community upgrading projects (Nunbogu et al. 2018).

Conclusion

Cities in Sub-Saharan Africa have faced the challenge of overcoming colonial legacies in urban planning – in theory and practice. Colonial-era planning often led to intense duality in urban forms and societies that has proven stubborn to undo in the post-colonial period. At the same time, colonial regimes often failed in efforts to enframe African cities and compartmentalize African urban lives. Urban Africans often worked to reframe the colonial order. Yet independent governments often replicated colonial planning tactics or at the very least failed to incorporate the reframing tactics of the formerly colonized. Structural adjustment and weak capacity for state-led implementation of formal planning furthered the sense that urban Africans were, essentially, on their own.

The consequent failings of formal planning have gone hand in glove with the informalization of urban development in most cities of the region, including those upon which this chapter has focused. Informal areas vary dramatically between cities in Ghana, Kenya, and Tanzania, but they tend to showcase innovative alternatives to formal planning. There are challenges with championing such informal planning, such as the frequency with which indigenous organizations, rogue officials, illegal cartels, or even organized crime operations dominate land and service provision systems in ways that run counter to the enhancement of rights to the city or deeper democratization. The current era of an African "renaissance" has had notable limitations, but it has opened up some possibilities for scaling up local, community-led planning in cities like Accra, Dar es Salaam and Nairobi. The way forward clearly lies in continuing the efforts to integrate small-scale innovations and successes at the community level with city-wide and national-scale urban planning frameworks.

Alternative planning history and theory timeline

Period	Planning history and theory canon	Alternative timeline: colonized peoples [Anglophone Sub-Saharan Africa]
Late 1800s–1914	Birth of planning	The establishment of colonial urban control Scramble for Africa, concessionary companies, dual cities, Indigenous placemaking
1918–1945	Formalization of planning	The formalization of colonial planning "Tropical" planning, building control, Garden Cities, segregation
1945–1965	Growth of planning	Planning and movements for independence Second colonial occupation, enframing, master planning, movements for independence
1965–1980	Midlife crisis of planning	Planning in the early independence period Ujamaa, Cold War rivalries, public housing, decolonization, authoritarianism, development
1980–2000	Maturation of planning	Urban planning amid structural adjustment Structural adjustment programs, urban service provision, sectoral planning, liberalization, informal settlements
2000–present	New planning crisis	Crises and opportunity in Africa's "renaissance" era "New Scramble for Africa," satellite cities, democratization, activism, deprivation, precarity

Case study: Kibera Public Space Project

The Kibera Public Space Project is a collaboration of the Kounkuey Design Initiative (KDI) and six communities in the Kibera informal settlement in Nairobi. KDI is an international planning/engineering non-profit partnership, registered in California, which began in 2006 with its work in Kibera. It has since developed similar projects in Haiti, Ghana, Morocco, and – using the insights from its work in these Global South contexts – in a trailer park in the Coachella Valley in California. KDI collaborates with local partners at the grassroots to create what it terms "productive public spaces" out of "formerly unusable and sometimes unsafe areas" in poor urban communities (see: www.kounkuey.org). They are "working to generate radical alternatives to slum conditions" in Kibera (Odbert and Mulligan 2015, 177; Mulligan et al. 2020). In contrast with most of the hundreds of well-intentioned projects in Kibera, KDI's work combines the expertise of technically trained planners with a keen understanding of the past of the place, a rich collaborative conception of both the physical and the spiritual dimensions of the Kibera landscape, an awareness of the artistic and creative capacities of residents, and, obviously, engagement with the complex and nuanced grassroots of Kibera neighborhoods. They aim for "true agency in the decision-making process" for a wide net of partners and stakeholders, avoiding oversimplification and disenfranchisement (Odbert and Mulligan 2015, 178). The evidence suggests that their work has "mobilized a network of Kibera residents with increased resilience and capacity to effect change" (Odbert and Mulligan 2015, 189). Mulligan et al. (2020) document the tangible achievements of the Kibera Public Space Project in the provision of community sewer infrastructure, gardens, tree-planting, and rainwater-harvesting capacity. The participants continue to ask the hard question of "why … slum conditions persist given significant investments and upgrading efforts" in places like Kibera, and they evaluate the project "not in terms of success or failure, but in terms of new insights" that can incrementally change the "urban realities of growing cities like Nairobi" (Odbert and Mulligan 2015, 177 and 192).

Note

1 For instance, the Low-Cost Housing Programme established in 1972 funded the development of several housing estates, including Dansoman Estates in Accra, and Ashaiman, Sakumono, and Manhean estates in Tema. The Bank for Housing and Construction was also created in 1973 to support housing and construction in general, while the Public Servants Housing Loans Scheme and the Armed Forces Mortgage Loans Scheme were also created to support housing provision for public servants and the military respectively (Adarkwa 2012).

References

Adarkwa, K. K. 2012. "The Changing Face of Ghanaian Towns." *African Review of Economics and Finance* 4, no. 1, 1–29.

Adjaye, David. 2011. *Adjaye Africa Architecture: A Photographic Survey of Metropolitan Architecture*, edited by Peter Allison. London: Thames & Hudson.

Alexander, Linda. 1983. "European Planning Ideology in Tanzania." *Habitat International* 7, nos. 1/2: 17–36.

Armstrong, Allen. 1987. "Master Plans for Dar es Salaam, Tanzania." *Habitat International* 11, no. 2: 133–146.

Banyikwa, William. 1989. "Effects of Insensitivity in Planning Land for Urban Development in Tanzania: The Case of Dar es Salaam." *Journal of Eastern African Research and Development* 19: 83–94.

Berman, Bruce. 1990. Control and Crisis in Colonial Kenya: the Dialectics of Domination. Athens, Ohio: Ohio University Press.

Bigon, Liora, and Yossi Katz. 2016. *Garden Cities and Colonial Planning: Transnationality and Urban Ideas in Africa and Palestine.* Manchester: Manchester University Press.

Bissell, William. 2011. *Urban Design, Chaos and Colonial Power in Zanzibar.* Bloomington, IN: Indiana University Press.

Brennan, James, and Andrew Burton. 2007. "The Emerging Metropolis: A History of Dar es Salaam, circa 1862–2000." In *Dar es Salaam: Histories from an Emerging Metropolis*, edited by James Brennan, Andrew Burton, and Yussuf Lawi, 13–75. Dar es Salaam: Mkuki na Nyota Publishing.

Briggs, John, and Ian Yeboah. 2001. "Structural Adjustment and the Contemporary Sub-Saharan African City." *Area* 33, no. 1: 18–26.

Burra, Marco. 2004. "Land Use Planning and Governance in Dar es Salaam: A Case Study from Tanzania." In *Reconsidering Informality: Perspectives from Urban Africa*, edited by Karen Hansen and Mariken Vaa, 143–157. Uppsala, Sweden: Nordic Africa Institute.

Carmody, Padraig. 2011. *The New Scramble for Africa*. Malden, MA: Polity Press.

Charton-Bigot, Helene. 2010. "Preface." In *Nairobi Today: The Paradox of a Fragmented City*, edited by Helene Charton-Bigot and Deyssi Rodrigues-Torres, pp. ix–xii. Dar es Salaam: Mkuki na Nyota Publishers.

Cohen, Andrew. 1959. *British Policy in Changing Africa*. Evanston, IL: Northwestern University Press.

Cooper, Frederick. 1987. *On the African Waterfront: Urban Disorder and the Transformation of Work in Colonial Mombasa*. New Haven: Yale University Press.

Coquery-Vidrovitch, Catherine. 1991. "The Process of Urbanization in Africa." *African Studies Review* 34, no. 1: 1–98.

Cristofaro, Domenico. 2020. "The Birth of a Town: Indigenous Planning and Colonial Intervention in Bolgatanga, Northern Territories of the Gold Coast." In *Routledge Handbook of Urban Planning in Africa*, edited by Carlos Nunes Silva, 15–29. New York: Routledge.

Dar es Salaam City Commission. 1999. *Strategic Urban Development Planning Framework*. Dar es Salaam: Dar es Salaam City Commission.

Dickson, K. B. 1969. *A Historical Geography of Ghana*. London: Cambridge University Press.

Fält, L. 2016. "From Shacks to Skyscrapers: Multiple Spatial Rationalities and Urban Transformation in Accra, Ghana." *Urban Forum* 27: 465–486.

Glassman, Jonathon. 2011. *War of Words, War of Stones: Racial Thought and Violence in Colonial Zanzibar*. Bloomington, IN: Indiana University Press.

GoDown Arts Centre and Kwani Trust. 2009. *Kenya Burning: Mgogoro baada ya Uchaguzi 2007 [the Trouble after the 2007 Election]*. Nairobi: The GoDown Arts Centre and Kwani Trust.

Grant, Richard. 2005. "The Emergence of Gated Communities in a West African Context: Evidence From Greater Accra, Ghana." *Urban Geography* 26, no. 8: 661–683.

Grant, Richard, and Yankson, Paul. 2003 "Accra: City Profile" *Cities* 20, no. 1: 65–74.

Hake, Andrew. 1977. *African Metropolis: Nairobi's Self-Help City*. London: Chattus & Windus.

Home, Robert. 1997. *Of Planting and Planning: The Making of British Colonial Cities*. London: Spon.

Huchzermeyer, Marie. 2011. *Tenement Cities: From 19th Century Berlin to 21st Century Nairobi*. Trenton, NJ: Africa World Press.

Huxley, Elspeth. 1948. "Review: Nairobi: Master Plan for a Colonial Capital, by L. W Thornton White, L. Silberman and P. R. Anderson." *African Affairs* 47, no. 189: 251–252.

Kequan, Qian. 1982. *The 1982 Zanzibar Town Master Plan*. Zanzibar: Revolutionary Government of Zanzibar.

Lanchester, Henry. 1923. *Zanzibar: A Study in Tropical Town Planning*. Cheltenham: Burrow.

Low, D. Anthony, and John Lonsdale. 1976. "Introduction: Towards the New Order 1945–1963." *History of East Africa*, volume 3, edited by D. A. Low and Alison Smith, 1–63. Oxford: Clarendon Press.

Mboya, Joy. 2014. "*Nai ni Who? (Who is Nairobi?)*: Collective Urban Vision Development." In *Visionary Urban Africa: Built Environment and Cultural Spaces for Democracy*, edited by Centre for Fine Arts, Brussels (BOZAR), 66–71. Brussels: BOZAR.

McQuillan, Aidan, and Royce Lanier. 1984. "Urban Upgrading and Historic Preservation: An Integrated Development Plan for Zanzibar's Old Stone Town." *Habitat International* 8, no. 2: 43–59.

Médard, Claire. 2010. "City Planning in Nairobi: The Stakes, the People, the Sidetracking." In *Nairobi Today: The Paradox of a Fragmented City*, edited by Helene Charton-Bigot and Dayssi Rodrigues-Torres, 25–60. Dar es Salaam: Mkuki na Nyota Publishers.

Mitchell, Timothy. 1988. *Colonizing Egypt*. Cambridge: Cambridge University Press.

Morton, David. 2019. *Age of Concrete: Housing and the Shape of Aspiration in the Capital of Mozambique*. Athens, OH: Ohio University Press.

Mulligan, Joe, Vera Bukachi, Jack Campbell Clause, Rosie Jewell, Franklin Kirimi, and Chilina Odbert. 2020. "Hybrid Infrastructures, Hybrid Governance: New Evidence from Nairobi (Kenya) on Green-Blue-Grey Infrastructure in Informal Settlements. *Anthropocene* 29: 1–17.

Mutisya, Emmanuel, and Masaru Yarime. 2011. "Understanding the Grassroots Dynamics of Slums in Nairobi: The Dilemma of Kibera Informal Settlements." *International Transaction Journal of Engineering, Management and Applied Sciences and Technology* 2, no. 2: 197–213.

Myers, Garth. 2016. *Urban Environments in Africa: A Critical Analysis of Environmental Politics*. Bristol, UK: Policy Press.

Myers, Garth. 2015. "A World-Class City Region? Envisioning the Nairobi of 2030." *American Behavioral Scientist* 59, no. 3: 328–46.

Myers, Garth. 2011. *African Cities: Alternative Visions of Urban Theory and Practice*. London: Zed Books.

Myers, Garth 2005. *Disposable Cities: Garbage, Governance and Sustainable Development in Urban Africa*. Aldershot, UK: Ashgate.

Myers, Garth. 2003. *Verandahs of Power: Colonialism and Space in Urban Africa*. Syracuse, NY: Syracuse University Press.

Ngau, Peter. 2013. *For Town and Country: A New Approach to Urban Planning in Kenya. Africa Research Institute, Policy Voices Paper Series*. London: Africa Research Institute.

Ngware, Suleiman, and J. M. Lusugga Kironde. 2000. *Urbanising Tanzania: Issues, Initiatives, and Priorities*. Dar es Salaam: Dar es Salaam University Press.

Njeru, Jeremia. 2010. "Defying Democratization and Environmental Protection in Kenya: The Case of Karura Forest Reserve in Nairobi." *Political Geography* 29, no. 6: 333–342.

Nunbogu, A. M., P. I. Korah, P. B. Cobbinah, and M. Poku-Boansi. 2018 "Doing It 'Ourselves': Civic Initiative and Self-Governance in Spatial Planning." *Cities* 74: 32–41.

Nuttall, Sarah, and Achille Mbembe. 2008. *Johannesburg: The Elusive Metropolis*. Durham, NC: Duke University Press.

Obudho, Robert. 1983. *Urbanization in Kenya: A Bottom-Up Approach to Development Planning*. Lanham, MD: University Press of America.

Obudho, Robert. 1982. *Urbanization and Development Planning in Kenya*. Nairobi: Kenya Literature Bureau.

O'Connor, Anthony. 1983. *The African City*. London: Hutchinson University Library.

Odbert, Chilina, and Joe Mulligan. 2015. "The Kibera Public Space Project: Participation, Integration, and Networked Change." In *Now Urbanism: The Future City Is Here*, edited by J. Hou, B. Spencer, T. Way, and K. Yocom, 177–192. New York, NY: Routledge.

Owuor, Samuel, and Teresa Mbatia. 2012. "Nairobi." In *Capital Cities in Africa: Power and Powerlessness*, edited by Stephen Bekker and Goran Therborn, 120–140. Cape Town: HSRC Press.

Pieterse, Edgar. 2010. "Cityness and African Urban Development." *Urban Forum* 21: 205–219.

Pieterse, Edgar, and Susan Parnell. 2014. "Africa's Urban Revolution in Context." In *Africa's Urban Revolution*, edited by Susan Parnell and Edgar Pieterse, 1–17. London: Zed Books.

Republic of Kenya. 2008. *Nairobi Metro 2030: A World Class African Metropolis*. Nairobi: Ministry of Nairobi Metropolitan Development, Republic of Kenya.

Simone, Abdoumaliq. 2004. *For the City Yet to Come: Changing African Life in Four Cities*. Durham: Duke University Press.

Smith, Constance. 2019. *Nairobi in the Making: Landscapes of Time and Urban Belonging*. Rochester NY: Boydell & Brewer.

United Nations Habitat. 2013. *Unleashing the Economic Potential of Agglomeration in African Cities*. Nairobi: United Nations Habitat.

Watson, Vanessa. 2014. "African Urban Fantasies: Dreams or Nightmares?" *Environment and Urbanization* 26: 1–17.

White, L. W. T., L. Silberman and P. R. Anderson. 1948. *Nairobi: Master Plan for a Colonial Capital*. London: HMSO.

Suggestions for further study

• Acheampong, R. A. 2019. *Spatial Planning in Ghana: Origins, Contemporary Reforms and Practices, and New Perspectives*, The Urban Book Series, Springer Nature Switzerland, https://doi.org/10.1007/978-3-030-02011-8.

• African Centre for Cities, University of Cape Town. https://www.africancentre-forcities.net.

• Calas, Bernard (ed.) 2010. *From Dar es Salaam to Bongoland: Urban Mutations in Tanzania*. Dar es Salaam: Mkuki na Nyota Publishers.

• Center for Urban Research and Innovations, University of Nairobi. http://www.centreforurbaninnovations.org.

• Grant, Richard. 2009. *Globalizing City: The Urban and Economic Transformation of Accra, Ghana*, Syracuse, NY: Syracuse University Press.

INDEX

Printed in the United States
by Baker & Taylor Publisher Services